TO NATASHA
WITH GOOD WISHES FOR
"OUR" NEW PROJECT!

Construction Cost Management: Cost Engineering, Cost Controls & Controlled Bidding

by

Adek Apfelbaum

authorHOUSE™

1663 LIBERTY DRIVE, SUITE 200
BLOOMINGTON, INDIANA 47403
(800) 839-8640
WWW.AUTHORHOUSE.COM

First published by AuthorHouse 11/16/05

ISBN: 1-4208-7140-4 (hc)

Library of Congress Control Number: 2005906645

Printed in the United States of America
Bloomington, Indiana

This book is printed on acid-free paper.

DEDICATED TO THE MEMORY OF

Doris Jane Apfelbaum

whose help would have made these pages more informative,
better organized and more elegantly expressed.
She was an exceptional lady whose
presence and assistance are sorely missed.

ACKNOWLEDGMENTS

My thanks goes to Charlene Smith, a freelance editor, who kept me from diverting from the main theme, who was able to decipher my scribbles and put together a readable volume. She was critical, but not too critical, editorially strong, but allowing me to use my own language. How she did all that was required without bruising my writer's ego is still an unsolved mystery.

My special thanks go to my son, Scott Apfelbaum, who in his capacity as the Director of Operations of a major university agreed to "field test" some applicable procedures, especially the process of controlled bidding. As in the case of Prada's Jil Sander project, it is estimated that the benefit to the owner/sponsor using this format saved 18% of the initially allocated/anticipated funds.

A. A.

OTHER BOOKS, PAMPLETS AND WRITINGS BY THIS AUTHOR

- *Cost Engineering and Aesthetics: A Guide for the Design/Construction Process*, copyright © 2005

- *Construction Cost Management: A Guide to Cost Engineering*, copyright © 2002

- *A Guide to Estimating, Cost Engineering and Cost Reduction Engineering Analysis (CREA)*, copyright © 1998

- *An Introduction to ID/C CREA*, an integrated design, cost and constructability analysis procedure, copyright © 1998

- *Recognition of Value*, copyright © 1997

- *Construction Cost Reduction: What's In a Name?* as part of *Cost Engineering*, Vol. 38/No. 4, copyright © April 1996

- *An Introduction to Cost Reduction Engineering Analyses*, copyright © 1996

Contents

FOREWORD

This recitation consists of two (2) parts. The first part (chapters I-IX) enumerates all requirements a developer/ institution should impose on its contractors and subcontractors. The second part (chapters X-XIV) is intended to serve as a cost awareness guide for an in-house project management team. Accordingly, all contracts executed by a developer with its construction managers, contractors and subcontractors, may make reference to this document and this document should be made part of it. To avoid confusion, the second part may not be included therein (the contract writer will make that decision). It is deemed useful, however, to have the contractor(s) obtain a full picture of the developer's/institution's makeup.

A developer/sponsor or owner/builder should be very cost conscious due to the large sums spent on his projects, some with complex designs. Accordingly, and to impress its many vendors and contracting entities to be like-minded, the owner/sponsor should establish or hire a cost engineering and technical evaluation section (CETES). This section is to be charged with executing pre-construction services (PC/S), cost reduction, cost engineering (CREA), budget preparation, change order evaluation and all cost control functions. As such, a typical department is staffed with estimators, cost engineers and certified cost reduction engineers (CCREs). Such a group would outline certain standards that will be imposed on all of its general contractors, construction managers, vendors and their subcontractors. Accordingly, this writing, especially the first part, should be made part of, and included in, all contracts and/or purchase orders issued by an owner entity. The underlying purpose of this writing is to make all entities dealing with an owner/sponsor very cost aware and prudent. An owner's funds should be available for good design, good detailing, excellent workmanship and high-caliber aesthetics, but not for wasted work, wasted management and costly duplications and mistakes. An owner should inform all involved that he intends to ensure that every dollar is properly spent and that he obtains a dollar's worth of work for every dollar spent. An owner should wish to have his contractors take advantage of his CETES department's technical skills and expertise to assure that together, the developer/owner and its construction partners, constantly look for and institute three most important ingredients in its building program: savings, savings and savings. A cost department or an out-sourced group look for it, strive for it and work hard for its realization. Contractors who can work within this premise should be rewarded; those who cannot or wish not to abide by this philosophy should be excluded.

I

PREPARATION FOR THE DESIGN/ CONSTRUCTION PROCESS

Before an owner/sponsor can effectively and economically realize his program and wish list, before he can even start discussions with his dream designer, he must establish cost consciousness and productivity evaluation capabilities. This can be accomplished by outsourcing the pre-construction services (PC/S) or setting up an in-house cost engineering and technical evaluation section (CETES).

An owner/sponsor, therefore, should institute pre-design and pre-bid conditions on any new project. Its in-house or out-sourced cost engineering and technical evaluation section should direct this function. The principles and methods used are to follow this chapter, which previously appeared as a separate document for Prada's construction activities and copyrighted by this author.

A prototype invitation, demand/outline and controlled bidding procedures for designers and potential bidders to be invited are to follow as a primer, which can be used as an implementation document. This will help to set the ground rules to avoid disappointments and delays.

PC/S, whether outsourced or in-house, will help set ground rules, which should bring the entire pre-construction process from sending the wrong signals that cause misunderstandings, misconceptions and, above all, miss-spending of funds. The reason many dream designer teams are not very prolific is their unfamiliarity with, or unwillingness to recognize the economic realities as they apply to the design/construction process. One entity, a cost control group, must be empowered to keep "the hot air balloon from floating away." A celebrity design team can be harnessed to produce a good design without the usual built-in bidding and money-wasting processes. Up-front planning and setting ground rules for economic attitudes will produce non-inflated results and, thus, good value for the money being spent.

— This page for annotations —

II

REQUIREMENTS FOR PARTICIPATION IN THE DESIGN/CONSTRUCTION PROCESS AND CONTROLLED BIDDING: A PRIMER

Note: This chapter, when treated as a separate document, may be reproduced for educational purposes.

An invitation:

You are hereby invited to participate
in a design and construction venture,
contemplated by the undersigned.
Please contact this office for time and location
of a pre-services orientation session
to take place soon, at which time
additional information and participation
requirements will be issued.
The pending project(s), as we visualize it,
fits well into your framework
and, therefore, we bring it to your attention.

With kindest regards,

Name_____
Company_____
Telephone Number_____

INTRODUCTION:

This institution is now in the process of establishing real estate participation and development in your working area. Toward that goal, we wish to establish a good working relationship with individual team members, who could participate with us on an "as need" basis. To fit into this total image, we expect you to abide by certain standards that have been established to assure that all participants, through their individual effort, create a synergy effect to obtain a better end product. A certain amount of (superfluous) interaction between members of the total team will be required. Further, basic conditions cited will become part of any agreement, if and when such an agreement is consummated. Such formal ground rules and pre-accepted conditions have proven productive in the past. We hope that this abridged version works well for both this institution and all participants.

REQUIREMENTS:

The basic requirements have been divided into two parts: the "Design and Pre-Construction" activities and the "Bidding and Construction" functions.

I.   The Design and Pre-Construction Period:

●   In addition to standard practices of the industry and the rules of conduct, as demanded by the A.I.A., this institution anticipates that before a design team is chosen:

A professional design team to be considered will be clear in its outline proposal to do the work, that it will encompass ALL work within its discipline, and that its work will be specific to the project in question. It will automatically be assumed that all insurance is included in any proposal and that each potential participant accepts the standard A.I.A. arbitration clause.

- It should be understood that this institution/developer of this project will accept only complete package proposals, i.e., a design team must be integrated under a one-umbrella entity, preferably the architect, to avoid having to deal with several subcomponent designers.

- The designer may provide his own contract form, or that of the A.I.A., which will be reviewed by the developer's pre-construction team before it becomes final.

- It is understood that the design team is willing to work in concert with the developer's pre-construction services provider and has allocate time in its total cost profile for cost engineering activities that will be conducted and paid for by the developer.

- The design team, at no additional cost, must be willing to provide a minimal amount of time for orientation meetings, exploratory discussions and other pre-assignment activities. Designer's cost for these activities is to be limited to a maximum of

$_____ or $_____. "Major" presentations will be charged to the developer at a predetermined hourly rate.

- The design team's proposal must include any and all out-of-pocket expenditures, such as material, transportation, duplication, and computer and research time.

- It should be understood that the design team, which will be chosen to develop economically viable designs, as per a developer's input, will assign rights to the use and reuse of that design to the developer.

II. The Pre-Bidding and Pre-Construction Activities (Controlled Bidding):

- The potential contractors, subcontractors, CMs and prefab manufactures will be requested to bid in two stages. For the first stage, they will be required to submit a budget price within a low and a high limit (established by the owner/sponsor), based on preliminary information available at the time. For the second stage, the two lowest bidders will be requested to participate in cost reduction sessions and only thereafter will they be asked to submit a final price.

- Proposals will be accepted only if they follow the format of the publication, *Construction Cost Management: A Guide to Cost Engineering*, First Books Library, pp. 70-82. Should this format prove too cumbersome for some smaller contractors, the developer my wave some or all of the requirements.

- The bidders will be required to provide evidence that they are bondable or have access to assets equal to 10% of the bid amount.

- The developer can use a simple two-page contract form, or utilize the standard A.I.A. contract, and will specify a lump sum not to exceed format with an allowance for general conditions.

- If a GMP (guaranteed maximum price) format is chosen, the developer will make all payments directly to the supplier in the contractor's behalf and will reimburse the contractor for general condition expenditures for the prior month.

- The developer will provide a "good faith" deposit and will pay 90% of the value of work in place at each requisition period.

- Insurance certificates will be mandatory.

- Owner's representation and the construction management format will follow the aforementioned publication.

- The pre-construction services (PC/S) should be performed in-house or, at least, under in-house direction. To assure that contractors perceive a project as a viable economic structure, they must be convinced that the owner knows its intrinsic worth, that he understands the constructability implications and that he has a good feel for the prevailing market conditions. Giving a contractor/CM an outer-limit estimate, prepared as part of the pre-construction process, establishes such a perception. This outer-limit estimate should be coupled with verbal orientations to reinforce the owner's control and demonstrate his involvement.

PRE-BID COSTING/ ORIENTATION MEETING/ WORK SESSION

After the basic requirements have been formulated by the designer(s) and before any project is presented to new GCs or CMs, the owner/sponsor will invite each for a pre-bid work session/orientation meeting. At this meeting, the cost department will:

- Elaborate on and explain the owner's wish list.

- Explain the design intent.

- Provide a quantity takeoff and a list of the owner's furnished equipment.

- Provide cost/productivity guidance.

- Assist the construction entities in evaluating the design intent by virtue of laying out cost reduction potentials.

- Look to the (sub)contractor(s) for cost reduction ideas and substitution suggestions, and help in the evaluation of the same.

- Together with the contractors and their suggestions, the total team should establish outside limits for the envisioned scope.

At this orientation session, the owner should request all contractors to participate in the PC/S (pre-construction services), which the owner's department should conduct, based on this primer on the topic. This participation will be conducted on a voluntary basis to help the contractors get a true picture of the intrinsic worth of the work, at no cost to the owner/sponsor. The owner's cost department should lead such brainstorming sessions and the contractors should be requested to send their designated PMs for on-the-spot consultation.

— This page for annotations —

III

PC/S,
PRE-CONSTRUCTION SERVICES:

WHAT IS IT?

WHO SUPPLIES IT?

WHO NEEDS IT?

WHAT DOES IT DO FOR A PROJECT?

HOW MUCH DOES IT COST?

DOES IT SAVE MONEY?

In the construction industry where there are many experts, every expert voluntarily admits to providing pre-construction services (PC/S). Yet, these services range from a simple acknowledgment that it exists, to a budget estimate, to a proposal to do the work. Few standards exist and fewer yet are GCs and CMs who truly have the skills, the full understanding or the interest in providing a client with the full services under the general heading of PC/S. Those GCs or CMs, who want and try to do more for their clients, simply assign PC/S duties to an estimating group. Others even rename their estimating department to imply that they offer PC/S as a matter of routine.

To establish a baseline of activities to be requested, to provide some semblance of order and to provide an outline of services, this abbreviated outline has been provided herein.

PC/S can be defined as any and all assistance given by an individual or a group versed in construction industry activities before the plans are committed for construction. Out of necessity, this assistance must include:

- wish list analyses
- conceptual/budget estimates
- constructability reviews
- system analyses
- market analyses
- packaging suggestions
- cost reduction suggestions
- life-cycle costing reviews

- design coordination assistance/conflict resolutions
- cost to detailing (on-the-spot) interpretations
- design/drafting excesses reviews

The design team, as the owner's/sponsor's agent, must assure that the entity assigned to provide PC/S covers a complete scope. The interaction with the PC/S provider is mostly with the design team and eventually with the construction team. For this reason, pre-construction services were sometimes known as design/construction (consulting) services, not to be confused with the owner's representative functions.

Traditionally, pre-construction services (in its many complete and incomplete forms) have been provided, and sometimes are still provided, by a designated GC or CM. This system has many built-in flaws, which are detrimental to both designer and owner. The GC/CM, under this system, is given the opportunity to lock himself into the project, eliminating any chance for competitive, free enterprise advantages. Further, the GC/CM has one major overriding interest and that is to protect his guaranteed cost, and not to help design the most cost effective project. Of course, exceptions do exist.

More recently, independent consultants have supplied PC/S, first, by estimating services and, more recently, by cost reduction engineers.[1] The use of an

---

[1] *Construction Cost Reduction: What's In a Name?* as part of *Cost Engineering* by Adek Apfelbaum, Vol. 38/No.4, copyright © April 1996.

independent group keeps the potential GCs and CMs on their toes and the owner/sponsor uncommitted during the assistance/planning phase to keep the value of his project at a maximum.

Any project, which requires more than a casual sketch to convey the owner's wish list, should have a PC/S practitioner or team on board. This can take the form of an independent consultant, a design team practicing integrated design/cost analysis procedure (ID/C CREA) or, only under special conditions, by a designated GC/CM who employs an independent outsider to keep himself honest. By implication, any major project, or any project expected to cost more than one million dollars, should have an effective and the most efficient control system in place. The one that looks out for the owner's interests and produces the most cost effective project is the independent practitioner who is also a good certified cost reduction engineer (CCRE). Cost reduction, more than anything else, is the most potent ingredient in a good PC/S program.

The owner's potential contractors and subcontractors should be requested to participate from time to time as may be deemed beneficial to the project. The cost for this participation is to be part of the contractor's bidding cost and each contractor is urged to include an amount for such participation in his bid preparation. The owner should only absorb the cost of his own personnel. Non-successful bidders will not have a claim against the owner for this intermittent service being requested as part of the contractor's bidding process.

PC/S is not only highly desirable; in many cases it is a necessity. PC/S becomes an important component of the entire design/construction process when a project is highly complex, when public-type bidding is not an option, and when the owner/sponsor wishes to apply most available funds to the construction (hard costs) and not contract documentation. In addition to saving money (by virtue of CREA), PC/S eliminates or reduces the need for many contingencies and change orders. It also reduces incidents of false budget expectations and avoids construction delays.

PC/S is a component of the total cost when applied directly (if actively applied) or indirectly (if resultant consequences run their course). Therefore, it should be controlled by allocating funds for a formal PC/S program. That amount averages 7% to 9% of the architectural/ engineering fee. It could be frozen into a lump sum, if such services are subcontracted, or it can be provided for the owner and contractor's benefit, at the owner's cost.

One should beware of free PC/S, usually offered by a GC or CM, for one gets nothing for nothing. It is certain that the GC/CM will make up the cost in some way. The best advice one can offer is for every owner/sponsor to employ directly a PC/S group and assign its duties to the design team to make full use of the group's abilities.

As pointed out heretofore, the most important component of a PC/S program is CREA (cost reduction engineering analysis). Another important component is minimizing change orders and unforeseen costs. CREA has been proven[2] to save 4% to 18% of the construction costs. Change orders, more often than not, cost the owner/sponsor an additional 8% to 18%. If only portions of these amounts are saved, a realistic gain may be substantial. These and other known probabilities point to a realistic expectation of 5% to 18% savings if a formal PC/S program is implemented.

---

[2] *An Introduction to Cost Reduction Engineering Analyses* by Adek Apfelbaum, copyright © 1996.

CREA,
COST REDUCTION ENGINEERING ANALYSES

Contractors invited to bid on an owner's/ sponsor's work are to be informed herewith that the owner's cost control section will conduct an ongoing cost reduction program, as described hereafter.

Contractors should also be urged to take the potential cost reduction effort into consideration when estimating their price for guaranteeing a maximum, not to exceed GMP (guaranteed maximum price) bid or a lump sum cost submission. Their intermittent participation in this process will be requested. Thus, their input should likewise be respected as a force in lowering projected costs. It should be understood that the arbiter of what cost saving suggestions will or will not be accepted is vested with the design team and the owner's management.

At any time during the bid process and thereafter during the construction process, all contractors will be reminded repeatedly to highlight time and money intensive design elements and bring those findings to the attention of the cost control section. They will evaluate those elements on a life-cycle costing basis and will report their findings to the owner's management, who will make the decision to pursue or not to pursue these suggestions further. No design element is to be left unexamined if it has a hint of a potential cost saving possibility. Belt tightening is the new battle cry of the twenty-first century, and an owner/sponsor should do all he can to promote cost efficient but good design.

— This page for annotations —

IV

AN INTRODUCTION TO
COST REDUCTION ENGINEERING
ANALYSES, CREA:

WHAT IS IT?

WHERE DID IT ORIGINATE?

HOW IS IT APPLIED?

WHO BENEFITS FROM IT?

DOES IT HINDER/HELP THE
ARCHITECT/ ENGINEER TEAM?

DOES IT DICTATE DESIGN?

WHO PRACTICES ITS ART?

WHO IS A CCRE?

Value engineering (VE) was first developed as part of the manufacturing processes, when staggering costs required the use of innovative and new cost reduction methods for product developers to be competitive. General Electric was the first to develop and use value engineering. They systemized procedures with forms and questionnaires, which allowed any engineering team to follow a given path, come up with areas to be scrutinized and think out alternative solutions.

The formula devised was:

$$\text{Value} = \frac{\text{Function}}{\text{Cost}}$$

This formula was the basis for further value-engineering (VE) development, especially the reduction of "C" (cost).

In the construction industry, VE has been adopted and adapted from manufacturing over the past thirty years. First, there was VE, and then applied VE, and now CREA (cost reduction engineering analysis), all of which have evolved into a total cost management package commonly called EACC (engineering analysis and cost control). CREA is the latest major component in the evolutionary process. In construction where there may be as many as 2,000 different components, an orderly process of analyzing system cost is paramount.

An example is the movement of labor and equipment in an open site versus the access to an area within a busy city block adjoining two existing structures and bordering on a busy street. In short, construction takes place in an uncontrolled setting, totally unlike factory production for which an environment can be designed for optimum efficiency and for which classical VE was created.

Classical VE forces engineers to squeeze each job into a standard mold and, thus, overlook some major areas of opportunity for improvement. CREA, adopted by most practitioners of VE in the construction industry, treats each construction project as a unique problem requiring unique solutions. It is part of a total control concept. EACC is defined further on in this writing. There are obvious similarities, which are (and should be) used as guidelines for attempting a VE solution. However, a group or an individual with years of experience must attempt it, someone who apprenticed with another cost reduction engineer, who has thorough training in engineering and architecture and is a graduate engineer or architect. Unfortunately, there is no curriculum in any university that prepares future engineering students as CREA practitioners. To be commercially accepted, one must have a track record, and no investor is willing to serve as a guinea pig. Today the Society for the Advancement of Construction Cost Engineering (SACCE) is accepting the responsibility of certifying practitioners as certified cost reduction engineers (CCREs).

Classical value engineering is defined as "an organized, creative approach, which has for its purpose the efficient identification of unnecessary cost, i.e., costs which provide neither quality nor use, nor life, nor appearance, nor customer feature."[1] CREA, its successor for construction application, may be defined as "the orderly application of economics, consumerism and engineering to the process of architectural construction." By redefining functional requirements, a solution becomes evident, which may be better, cheaper and, thus, of more value, less cost or both in many cases. <u>CREA should not be confused with other cost reduction programs</u> that are narrower in scope and usually focus on one discipline only. For example, a structural engineer, who is an expert in tension structures, may change a structural system using his method, which requires less steel (Dr. Lev Zetlin used this method successfully to devise cable systems for hangers and other long-span structures). CREA is not only an art of substitution, but also the use of unorthodox approaches to solving problems that involve value (consumerism), function (engineering) and cost (economics). Traditions (prior practices) are the most obvious stumbling blocks to proper cost reduction. Anything produced by man may be improved and simplified. Sir Clive Sinclair demonstrated a good example of cost reduction. He reduced the number of chips needed for his home computer by devising a way to assign several tasks to the same chip, utilizing their spare waiting time and capacity. The result was a large V (value) by increasing F (function) and decreasing C (cost). When the function became insufficient, when his computer could no longer do what others could, no reduction in cost could hold the value at a level to interest consumers. A perfect example to prove that the formula has limitations:

---

[1]*Techniques of Value Analysis and Engineering* by Lawrence D. Miles, 3rd Edition, copyright © 1989.

$$V_{max} = \frac{F}{C_{min}}$$

Costs can approach, but never reach; zero and V (value) have limits.

As VE evolved, engineers looked to the construction industry as a perfect candidate for cost reduction application. Construction has all the ingredients to make value studies effective: many components, system groupings, considerable hand assembly, large fluctuation of cost and variable value assessments. Thus, CREA was born, defined as an orderly approach to budgeting, estimating, cost comparison and tracking of construction cost statistics. It is often referred to as the point where engineering, architecture and economics meet.

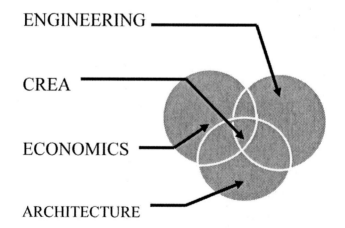

ENGINEERING

CREA

ECONOMICS

ARCHITECTURE

CREA is the one part of total EACC concept, which uses constructability reviews (How easily can the plan be put together?), estimating and budget preparation (How realistic is the total anticipated cost?), scheduling (Can the project be designed or built within the time allotted?), life-cycle costing (How much will the total system cost during its anticipated lifespan?), change order management (How many additions and deductions are needed to complete the project and who caused them?). CREA is the most crucial part of EACC. It is the commercial application and final evolution of traditional VE and concerns itself mainly with decreasing the C (cost) in the revised formula where E stands for esthetics.

$$V = \frac{F + E}{C}$$

F and E (function and esthetics), as used in this CREA formula, are left to the design team, unless blatantly abused.

Although not intended to be an instruction manual, the following will illustrate the approach CCREs (certified cost reduction engineers) take as it pertains to the construction industry. Every design has a potential for becoming better and less costly, no matter how careful and thorough the design team may be. This does not reflect on the negligence or lack of ability of the designers. It simply means new solutions can be found by redefining requirements and re-analyzing systems, and by combining or separating functions. Even after a cost reduction review has been conducted, a new review may find new and additional solutions. A second team may find even more solutions. The approach must be customized to each project. Since esthetics is a major component in architectural construction, no method is applicable in all cases. For this reason, the architect becomes a strong force in CREA. Also, because of the diversity, each practitioner has adopted his own unique method. CREA is, therefore, an unorthodox approach, made commercially acceptable by its practitioners who developed their own style and form, and is now standardized. Presently, there are few firms in the United States providing such service. Training is obtained usually in civil engineering and the practitioners' professional experience is in estimating and quantity surveying. There is an effort underway to persuade design firms to adopt an integrated design/construction process based on CREA practice.

Although CREA is only a component of EACC, it is the one that produces results in a project for the individual or group investing. It is best applied early in the design phase to be most effective. The earlier CREA is applied, the larger the potential impact.

CREA starts by analyzing the total picture, usually formulated as:

$$Ct = \sum D+F+P+(gc)+Ws+Wo+c+(lcc)$$

Where:

Ct	=	total cost of project
D	=	design costs
F	=	cost of financing
P	=	promotion
(gc)	=	direct costs, general conditions

Ws	=	subcontracted work
Wo	=	owner's work
c	=	unforeseen contingency
(lcc)	=	life-cycle costing

In most cases, the design cost cannot be changed. It is a function of intangibles (friendship, esthetics, preferences, name and reputation, expertise, etc.) and not subject to free-market fluctuation. Therefore, it is usually untouched by CREA reviewers. Other components must be compared to known costs for similar projects to assess their ratio to the total cost (Ct). Any system within each component in the formula, if not within the acceptable limits of the known ratio, becomes a prime candidate for CREA review. Any system, which by itself exceeds 20% of "Ws + Wo," is, likewise, a prime target for review. Since the design is not yet frozen at the stage of analysis, the design team is consulted to insure that the intent was properly interpreted and that the estimated cost for the component or system in question was properly assessed. If the system or component is indeed estimated properly, the CREA reviewer then analyzes the system with respect to the formula "$V = F + E/C$." Is the "F" (function) greater than required and can the project live with an "F-" (lower function) and a corresponding "C-" (lower cost), thus, retaining the same or greater "F/C" ratio and, thus, the same or greater "V" (value)? Or will this "F-" change the system beyond requirements? An additional analysis would find alternative, less costly ways to provide the same function. That is, lower "C" (cost) by changing details and materials:

$$V+ \; = \; \frac{F + E}{C-}$$

The process is one of questioning systems, redefining the functions and trying alternative ways to accomplish the same at lower cost. Lowering "C" (cost) is the prime purpose of CREA. For example, the function of an air-conditioning system is to cool. It is not to own compressors, air handlers, diffusers, and structures to house and support them. The question for analysis is then: Was an unnecessary roof created for a compressor? Could several smaller units do the same job, or one large unit instead of smaller systems? Are there other units that do not need roof placement? The cost reduction team concentrates on the basic function and researches a solution that provides the same function without extraneous and peripheral construction. <u>A good cost reduction engineering analyst (CREA) never imposes his will or changes the architectural integrity of a project</u>. He provides alternate solutions that are acceptable to the design team and are sensitive to the design requirements. Never is esthetics ("E") questioned or even subject for review.

After each analysis, a design review is written giving the following:

- system description as is
- quantity involved
- new system description
- new quantity
- dollars saved

- life-cycle implication
- other alternatives

An engineering change proposal (ECP) is then prepared requesting formal acceptance of each change. This procedure can help the architect understand the process of re-definition and, thus, helps the ECP acceptance. After a give-and-take period, all changes are summarized and a final report issued. A new estimate follows showing that the ratio of each component to the total cost should be < or = than the accumulated databank norm with no more than a 5% to 10% deviation.

Few investors (individuals, corporations or organizations) realize that their investment in constructing any project can be maximized. Every investor knows that he must have an accountant, an architect, a structural engineer, etc. He does not always recognize the need for cost control and a team totally dedicated to this topic. Time is money! Unworkable details and overuse of materials, likewise, cost money. A CREA team can maximize each dollar spent. A good team does not impose design solutions. Esthetics is left to the architect. The economic execution of the esthetics, however, is best left to the project manager (PM), who becomes more effective when backed up by a CREA team. CREA or ID/C CREA (an integrated design cost constructability review procedure) is an adjunct to the total design/construction team, enhancing its capabilities. It is a synergetic process that makes the total effort more productive than otherwise possible.

Projects for cost reduction analysis, however, must meet certain criteria. A project should be above $500,000 and above the simplest form of construction to warrant employing a CREA or ID/C CREA team. Although every project can be a candidate for some savings, some low-cost, low-quality structures do not warrant the effort. How often a given design was utilized is not a function of saving potential. Even a cost-reduced design, after a given time, can be re-evaluated for new savings due to changed conditions and/or new developments.

In choosing to avail oneself of CREA, there are two possibilities: an independent CCRE consultant or a designer who employs an ID/C CREA system.

If a CREA consultant tells you that on a project similar to yours he suggested an X solution and that that X solution can be inspected to visualize how it appears, you have the wrong engineer. CREA is looking at and defining problems from new perspectives. If the solution worked once, it may not work again, or work as well on a new problem, as it did for a similar problem. <u>Each problem is unique and requires unique and customized approaches</u>.

A good CCRE is a non-conventional thinker. By definition and title, he has enough background to question many systems in order to arrive at a suggested solution. A CCRE does not have to know each system better than the consultant charged with the system's development/ design (the structural engineer, the elevator consultant, the HVAC engineer, etc.). However, he must know how each system integrates with other systems. This is where the practical approaches come into play. Theoretical know-how is important, but how each system intertwines with other systems is learned in the field. One must see, observe and remember. An engineer becomes a CCRE because of these qualities.

An employer of a CCRE should not question his ability to design a better elevator system than the elevator consultant, but rather question his understanding of the interrelation between the elevator and the structural system, the bulkheads, stairways, ramps, railings, shaft, subsoil water, access during construction, and other factors. At first glance, these factors may have nothing to do with the elevators and, thus, are often ignored by the elevator consultant. It is important to remember that a CCRE can do his work best if he is made a member of the design team and not be expected to work in a vacuum. The architect and his team's involvement are paramount for cost engineering to succeed.

The personal outlook of a cost reduction engineer is also very important. He must be able to co-exist with the designers without making them negligent or inferior. He, as a "spoke," is useless to the wheel unless he works in concert with other parts.

In short, a CCRE is an engineer, who has a good background in design and construction, is unconventional in his approach, not burdened by habit or imitation, and is certified by the SACCE (the Society for the Advancement of Construction Cost Engineering).

The bottom line of CCRE or CCCE (certified construction cost engineer) is <u>saving money</u>. Not much has been gained if that savings has to be spent for CREA activities rather than for doing the construction work. Therefore, the incentive for cost reduction application must be savings above and beyond the cost of CREA.

No developer should go through this exercise for the sake of academic curiosity. The savings must be real. Experience dictates that reduction engineered savings, if well conceived and executed, must be equal to a multiple of the cost of its application. An acceptable ratio is ten to one. That is, for every dollar spent on fees, the clients should reap ten dollars in savings. An experienced CREA team can increase this ratio to twenty to one. Thus, the fee paid should be directly related to the benefits gained. CCREs are fully aware of this phenomenon and, therefore, will tailor each fee to each project. If the project has a large potential for cost engineered savings, the fee will be sized accordingly. How much work is involved, how many reports are necessary, or how many man-hours are needed, have little to do with fair compensation. Most CCREs will charge in one of three ways:

- A flat fee of 0.5% to .75% of the construction total cost, exclusive of land, legal, architectural and other "soft" costs.

- A good faith payment (out-of-pocket reimbursement) equal to 0.5% of the cost, as above, plus a sharing of the savings, ranging from 18% to 40%.

- A combination of both.

It should be noted that under a flat percentage, the client should insist on freezing the percentage at the outset, converting the fee into a fixed lump sum (based on budgets established). This will eliminate any suspicion that a CCRE may not use his best efforts to reduce costs to the lowest level (thereby reducing his fee). Under the percentage of the savings category, the client must establish who will be the sole judge on the value of the savings. The project manager, the architect or the project attorneys are all likely impartial arbiters. Payment should be made on the same basis as those assigned to the architect.

To a large extent the fee levels depend on the CCRE's involvement. The fees cited are for full involvement working with the design team during the design process. For engineering review after the design has been completed, a CCRE is entitled to:

- 12% to 20% of the savings

or

- $2,000 to $3,000 per million of the budgeted construction cost

Most CCREs operate in such a way as to make their fee structure flexible. Beware of a CREA practitioner willing to take on an assignment at any cost and under any condition.

A CCRE has limitations on how many clients he can serve. A potential client should be aware of this. An overworked cost consultant cannot be very effective. However, because of its nature, cost engineering is not functionally dependent upon a large number of technically competent individuals. Unlike architecture, the size of the cost engineering office (number of employees) does not dictate the size of the project a firm can handle. A cost engineering office does not turn out volumes of drafting material or quantity takeoffs. Its end product is competent reporting, comparison and analyses. Thus, it is the quality of its personnel that governs its capacity rather than its size. A small cost engineering firm can easily service several major projects. An acceptable ratio is $100,000,000 worth of construction per cost engineering employee, per annum. A typical five-man cost engineering office, consisting of a senior engineer, one junior engineer, one economist, one editor and one secretary, can easily absorb projects budgeted at one-half billion dollars, without relying on additional outside personnel. This may be one single large project or several smaller ones. However, as the number of projects increase, the ratio of construction dollars to the firm size is not necessarily valid. Each project demands a minimum amount of attention regardless of its size or complexity.

A good guide is 1.5 projects per person on the staff. The theoretical firm cited could comfortably service one-half billion dollars worth of construction or eight projects with a sum total of less than $500 million. The rule, therefore, can be summarized as:

$$VP \text{ max} = S \times 10^7$$

or

$$NP \text{ max} = S \times 1.5$$

where:

- VP max = maximum volume of project in S
- S = size of firm, including secretarial and support staff
- NP max = maximum number of projects

The lower of the two values, "N" or "V," which fits this generalization, should be used as the governing factor. The value used for "N" can be rounded off to the nearest integer as was done in the example. For larger firms, factors can be added to the "S" component to compensate for the contributions made by each member of the team. As an example, additional secretaries do not increase capacity in the same proportion as additional engineers, architects and economists.

Many CCREs work as independent, self-employed individuals and, therefore, rely on part-time assistance, which they acquire as needed. Their capacity is a function of their energy. As a rule, each CCRE can handle simultaneously four to five projects.

Cost reduction engineering is not a miracle application of mumbo-jumbo magic. It is not a secret cure by a secret society member. A CCRE cannot pronounce an incantation and make all construction development problems go away. If the design/construction team is incompetent, disinterested or inexperienced, CREA cannot change the total equation. A badly conceived project will produce a bad design, which will be translated into a bad project. If, however, the team is reasonably expert, the basic information assembled is reliable and, if a realistic goal has been set, a CREA team can improve substantially the planning and execution. A CREA's influence far exceeds his seemingly minor role in the totality of efforts. Although the main purpose of cost reduction engineering is to improve the overall cost picture, one must recall the fundamental equation most cost engineers have indelibly imprinted in their professional thinking cap:

$$V = \frac{F + E}{C}$$

If costs cannot be lowered, the function should not be increased necessarily to obtain more value. A good cost engineer will also be familiar with proper detailing of architectural features, waterproofing, and, above all, methods of installation (i.e., field procedures). Thus, having a cost engineer on the team can help the architect obtain full potential from his staff (detailers, draftsmen). By consulting the CCRE, he can obtain much practical advice, which will minimize his research and eliminate the modification of his design. A simple procedural change many not save noticeable sums, but may later avoid field disputes, maintenance problems, and claims or worse.

This is especially true when new systems are being tried. For these reasons and other cost related realities, architects using ID/C CREA get preference by developers/investors in the private sector.

CCREs are trained in written and oral communication. In most cases, their reporting is superior to other team members and, thus, can be used as a whipping rod to improve the overall planning and scheduling. This reporting capability can (and should) be used as a psychological weapon to benefit the project. Elaborate reports when located on the negotiating table invariably intimidate the other side. Thus, bankers, mortgage brokers, tenants, suppliers, contractors and subcontractors are put at a negotiating disadvantage. Public institutions find the employment of cost engineers helpful when confronting potential donors/ sponsors. The institutions appear more efficient than they are, more serious, more thorough and, above all, more conscious of the funds being entrusted to them.

COMMON MISCONCEPTIONS

• Why do I need a CCRE if I hire the best design team?

A designer, even a good one (if the consultant is not good, he should not be on the project), is paid and expected to design the project to the best of his ability according to his subspecialty. He is interested in doing this in the most expeditious manner possible. He should take into account all variations and systems that could be applicable to the project. He is not required, nor is it in his interest, to worry about the interaction between his discipline and that of another consultant unless it is necessary. He cannot and is not trained to spend time on cost engineering analyses and "what if" situations that take into account multi-disciplinary systems affecting several consultants' work, unless the design firm uses ID/C CREA.

An example to illustrate this: The structural engineer is hired to design the best structural system for a particular structure, taking into account the size of the structure, site conditions, climate, architectural considerations, etc. If the air-conditioning consultant designs a system that has special load or beam-spacing requirements, the structural consultant is required to adapt his design accordingly. A cost engineer might suggest a modified method of hanging ducts, which would not change the HVAC system, but simplify and affect related disciplines (structural in this case). While causing the structural engineer to spend slightly more time and effort to design the structure, it may save either original expenditures or life-cycle costing, or both. This change may be beneficial to the owner and the other consultants, and be cheaper, even including the redesign cost, but would require the structural engineer to work a little harder. The engineer is not motivated, trained or expected to seek out such cost saving solutions on his own.

A cost reduction engineer (CCRE) spends all his time examining every system and subsystem to search for cost saving solutions. Many proposed changes do not produce savings for one reason or another and these are rejected during the course of analysis.

• Why do I need a CCRE if I have a project manager?

On governmental projects where cost reduction engineering is more commonplace than on private designs, project management companies frequently have CREA consultants on their staff. In most cases, however, even the best construction managers are concerned with ensuring that the design gets completed and that there is a smooth transition from design through the bidding process to successful construction. They spend their time with the consultants and the contractors, looking out for payments, sources, quantities, permits and procedures. Some project managers have additional functions, but none have the training that CCREs are required to possess. Even if they had such training, the analysis takes time and a detachment from the day-to-day pressures and conflicts of the project. A CCRE must be totally impartial as to which systems are to be analyzed, which consultants are affected, and which disciplines require adjustments. The construction manager's scope neither includes nor demands cost reduction analyses, which involve the comparison of every system with a theoretical "norm." Some project management firms employ the services of outside CCRE, which serves a subcontractor for the purpose of providing additional services to the owner. Some design firms use ID/C CREA in which no outside CREA is required.

SAMPLE APPLICATIONS

FACT(S)

On a given project, the total air-conditioning (A/C) cost equaled to 32% of the total estimated costs:

$$\frac{\text{Component cost}}{\text{Total cost}} = 0.32$$

Data kept by the CREA showed that this value for this project should be < or = to 19%.

A review indicated that several large compressors and cooling towers were being used. Because of many potential users of the system, multiple controls and re-heaters were being utilized and, because the roof was vaulted, new flat structures were being created to house the compressors.

ANALYSIS: DEFINING THE PROBLEM(S)

New roofs are costly and have no role in cooling. Compressors are commonly placed on the roof because of convention. The controls and re-heaters are a requirement (F). Large units were used because of the volume involved.

REQUIREMENT(S)

Each user needs A/C to suit his needs, but does not need complicated zoning. Each user is not interested in the size of the units and not interested in his neighbor's needs.

Although the system, as designed, had the required capacity, it had an F (function) greater than required. It needed additional structural work beyond its functional cost requirement. The total system is overvalued; V (value) larger than required.

SUGGESTIONS

Compressors need not be on the roof. Each user is better served having a separate unit. Therefore, create ground-level housing for small uniform compressors and run the refrigerant piping underground to each user.

FACT(S)

Israeli convention dictates that stairs be covered with pre-cast terrazzo treads and risers. On a high-tech building, a local architect designed a main stair, plus two-fire exit stairs at each end of a long wing.

ANALYSIS: DEFINING THE PROBLEM(S)

The function of the exit stairs was only for emergency use and, thus, will not receive much traffic. The pre-cast terrazzo treads were not functionally related to its use.

REQUIREMENT(S)

By law, an emergency exit is required and, therefore, stairs cannot be eliminated. However, elegance and appearance are not a functional requirement.

SUGGESTIONS

Omit pre-cast terrazzo treads, trowel concrete and use deck paint.

RESULT

With fourteen risers per floor, and two stairwells on an eight-story building, the savings were substantial for this portion of the work (pre-cast terrazzo treads). The savings did not impair the function of the building.

DEFINITIONS

A/E
Architect/engineer, design team.

CCRE and CCCE
A construction cost reduction engineer and a certified construction cost engineer.

Change Orders
A written document for changes to the project required, as a result of changed programs, program omissions or errors, field conditions (created or inherent), or an ECP (engineering change proposal) acceptance.

Clerk-of-the-Work
An engineer, architect or technician assigned by the CM to be the field liaison to the contractor and oversee the day-to-day workings of the contractor or subcontractors.

CM or PM
The construction manager or project manager appointed by the owner.

Constructability Review
A study of systems, subsystems, details and material choices to assess their buildability within the financial constraints of a given project and within the limits of the ability of local craftsmen and contractors. Suggestions for less-complex systems, or more easily assembled systems or detail, are part of this review.

Cost Reduction Engineering
The process of quantifying the cost of details or construction system, the suggestion of equal but less-costly systems or details, and the ability to forecast the cost of construction of details or systems used in construction. Working with the formula $V = F + E/C$, a cost reduction engineer concentrates on lowering "C to increase "V."

CPM
Critical path method of scheduling construction developed by the U.S. Department of Defense (also known as PERT, Program Evaluating and Review Technique).

CREA
Cost reduction engineering analyst/analyses

Detail Analysis
The investigation of architectural and structural details for constructability (ease of installation) and cost effectiveness (best price). The suggestion of alternative methods to arrive at the same visual impact is usually associated with these analyses.

EACC
Engineering analysis and cost control, usually referring to all aspects of cost and construction consultation during the design and construction process.

ECP
Engineering change proposal is a sketch or descriptive suggestion for cost reduction or system improvement/change.

Fast-Tracking
It is the acceleration of the design/construction process by overlapping some or many of the functions. The building is divided into components and some are given out for construction, while others are still being designed. Thus, the excavation, foundation and superstructure framing are being built, while the designers are still working out interiors, façade, mechanical systems, etc.

Life-Cycle Costing
The cost projection of a construction system during the "lifespan" of the structure (or system), including initial installation cost, maintenance and replacement.

Market Studies
It is the investigation into the availability of materials, contractors and subcontractors for a specific project for the purpose of targeting bid acceptance.

Packaging Study
It is the investigation into the breakup of a project into stages or components to obtain the best possible bids.

Total Maintenance Cost
This is the cost projection of a construction system for the "life" of its existence.

VECP
Value-engineer change proposal.

SUMMARY

Cost reduction engineering analyses (CREA) is an orderly attempt at finding new solutions to reduce the cost of anything being constructed, by redefining the requirements, and by finding alternate materials or methods (analyzing and redesigning entire systems or components). CREA is targeted only at the construction industry and involves redefining functions, analyzing systems and finding unorthodox ways to use systems or materials in a less-costly manner. It is the practical application of field experience and design know-how.

This chapter answers basic questions about CREA in construction:

1.   What is it?
2.   Where did it originate?
3.   How is it applied?
4.   Who benefits from it?
5.   Does it hinder or help the A/E team?
6.   Does it dictate design?
7.   Who practices its art?
8.   Who is a CREA?

— This page for annotations —

V

FAST-TRACK COST INTERPRETATION, ID/C CREA:

AN INTEGRATED DESIGN,

COST AND CONSTRUCTABILITY ANALYSIS PROCEDURE

THAT RESULTS IN COST REDUCTION AND BETTER DESIGN

Cost reduction, constructability reviews and total cost management during the design and pre-construction process should, and always have been, a must item on the minds of those who commission design assignments and those who manage the design process. Unfortunately, the topics have not yet been standardized and, as a result, not too many owners/sponsors or designers do more than pay lip service to the functions and the names that describe them. A design without full cost/constructability control is like a signed blank check.

Despite its emergence some thirty years ago, construction managers use pre-construction services (the term most often applied to controlling the design/budgeting process) whenever they are brought in during the design phase. Few, if any, design firms offer such services on a voluntary, self-imposed basis and fewer yet have the in-house capabilities to provide full cost management.

This writer has always been cost conscious of design, as it relates to functionality. As a result, he has adopted a way to integrate the services of cost reduction engineering analysts (CREAs), estimators, field engineers and risk analysts to evaluate the owner's wish list, alternate design solutions and the constructability of those solutions, in order to provide good value for money being spent. The result should produce:

- better design
- maximum functionality
- lowest possible cost
- maximum construction speed
- good value for every dollar spent

The system involves simple but logical steps that up to now have been disjointed, haphazardly applied and very often ignored.

Since value engineering was first introduced to the construction industry, there have been many attempts to improve and adapt cost reduction to the entire design/construction process. Developed by estimators, cost reduction engineering analysis (CREA) was first adopted by the United States Army Corps of Engineers for use on one of its own major projects. Eventually, senior estimators began using CREA whenever they were asked to interpret design/cost questions. Estimators, who by virtue of their experience had an early, clear visual conception of a planned project, were the logical originators of a new breed of cost analysts now practicing as CCREs. This new method of analyzing costs, with sensitivity toward the design intent, is the latest evolutionary stage in integrating designers' work with economic realities. It helps set and keep a realistic lid on costs, making the designers more effective and less prone to criticism by their clients. Under this system, the process takes on a predetermined course under the owner's control.

Upon identifying a potential project, an owner's dedicated CCRE indicates and explains the integrated system and requests basic information, including the owner's wish list. This information is then digested at a brainstorming session attended by the designers or chief designer and the CCRE. The result of this session produces a basic design concept, a budget and design/construction duration. A formal proposal/summary can then be formulated to show the full intent and the next step. If not accepted, this summary should be dissected and analyzed by the entire group so that future proposals/summaries can have appropriate parts of the proposal adjusted (cost, duration, design philosophy).

If accepted, this summary is brought before the design group to formalize a design concept, to establish how design intensive it must be to obtain the desired results, how the final wish list can, and should, be incorporated, and what method of construction is to be recommended (fast-tracking, straight bidding, construction management system, etc.). All decisions reached are then summarized by the CREA group and a report issued. During the design development, the total ID/C CREA group meets on a regular basis to protect against runaway costs and over design. At least twice during the design development, the ID/C CREA group prices the design in order to avoid surprises after bidding and costly embarrassing redesigns. In summary, it can be stated that ID/C CREA is a system/ procedure of checks and balances. The cost team checks and keeps in check the designers who may overstretch the budget, and the designers check and keep in check the CREAS who may want to destroy the design by over cost cutting.

In addition to those cited heretofore, the most obvious advantages are those the client or developer receives: better design. Freed from costing alternatives, the designers can concentrate on innovative and artistic expressions. The CCREs will do the rest. Other obvious advantages are:

- overall lower costs
- simplified contract scopes
- fewer change orders
- less litigation
- fewer overruns
- better client/designer relations
- maximum functionality
- ability to predict final expenditures
- on-the-spot design to cost evaluations

— This page for annotations —

VI

ESTIMATES[1]

CONSTRUCTION MANAGER'S ESTIMATING ACTIVITIES:

FUNCTIONS DEFINED

AND PURPOSE SUMMARIZED

---

[1] As part of the manuscript of *A Guide to Estimating, Cost Engineering and Cost Reduction Engineering Analysis (CREA)* by Adek Apfelbaum, copyright © 1998.

A GC's (general contractor's) and a CM's estimate requirements differ slightly due to the need they fulfill. This chapter concerns itself mainly with the CM's requirements.

CM's estimators/cost engineers must function as a (joint) unit for several reasons:

- to obtain higher productivity through synergy of the process.

- to be able to share the total workload.

- to avoid duplication of actions.

- to share in the combined know-how.

No individual should work in a vacuum nor create his own isolated domain.

Estimators or, more correctly, cost engineers, must be trained or self-trained to see the overall picture of their objective and purpose/intent, not just individual line items.

Estimators are used during pre-construction:

- to establish a budget

- to set a benchmark for cost negotiations and/ or cost reduction, if that becomes necessary

- to adjust a budget

- to confirm a MGP (maximum guaranteed price)/ GMP (guaranteed maximum price)

Estimators are used during and after construction:

- to review the budget

- to update the GMP, if such exists

- to review ECPs (engineering change proposals), PCOs (potential change orders), COs (change orders), or whatever term is being used to define increases or decreases to a CM's contract

Before any estimate is prepared, one must define what the purpose of the estimate is and how it will be used. Is the estimate to be an assessment of intrinsic value for further negotiation or is it to be used to adjust a scope or an agreement? Is the estimate to be used for acquiring additional work or to keep a contractor/subcontractor honest? Different purposes require different approaches. In general, estimates adhere to the following format:

- **Bidding Estimates or GMP estimates:**

Estimates must contain accurate quantities summarized after measurements and calculations are obtained instantaneously (without a worksheet listing dimensions) and rounded off to the nearest practical number. Unit prices used should take into account all peripheral work such as nuts, bolts, connections, welding, assessors, etc. Unit prices should not include general conditions, overhead and profit or any other soft costs. An estimate that is burdened with many line items or insignificant figures is no more accurate than an estimate that considers systems in total. All totals should be rounded off to the nearest significant number and then an estimator's judgment should be used to make the total larger or smaller in order to achieve a realistic number. Contractor's/subcontractor's confirmation is helpful, but not essential, and should only be used as a guide to the assessment. An estimator or someone who reviews estimates should never assume that a contractor's/subcontractor's estimate is more accurate than one's own. Unit pricing may come from experience, suppliers, past contracts or price books; a cost engineer, however, does not take these prices at face value, but must learn to interpret the source. For example, government unit costs tend to be based on previous payments, contracts or change orders, which are historically above market value. Without any adjustments, these unit costs can be misleading, driving estimated evaluations upward and out of sync with reality.

- **PCO/ECP Estimates or Assessments:**

They are prepared to check the intrinsic value of work segments and check contractor's legitimate needs and/or demands. They are used to provide a base for further negotiations of contract adjustments. In every case, the contractor requesting or being requested to do the work should provide actual quantities and unit prices. An estimate/assessment for these needs requires more of a feel of the contractor's intent and integrity and conditions of the affected work than the number the actual units provided. A contractor may mislead a reviewer (cost estimator) by listing insignificant line items on his estimate (more line items do not make a more accurate estimate). In this type of estimate, a statistical overview may be more helpful than a bean-count, micro-estimate. Contractor's units/quantities, after a cursory check, may be used to obtain a baseline value of the work invoiced. A new takeoff is required if the contractor is found padding either quantities or units. In any case, all review or assessment estimates should be concise, the source listed and only the major line items reported and summarized. The estimate and its backup should be made available to the change order negotiator. An estimator/cost engineer should be able to resolve scope and cost, but should never be called on to negotiate a change order. A good cost engineer does not automatically make a good negotiator. There exists a natural conflict of interest in those two functions. Above all, an estimate for a PCO evaluation should be viewed as if one's own expenditures were involved.

- **Special-Need Estimates:**

There are needs that must be evaluated on an indi-vidual basis. Such estimates must be adjusted accordingly using the rules listed heretofore, but summarized again for clarity:

- Establish the purpose before the estimating format is set.

- Focus only on the significant facts or quantities.

- Fewer line items make for a better estimate than a total "bean count" with many items.

- Soft costs should be listed separately and not part of unit costing.

- One should not automatically assume that all estimates prepared by contractors are inflated for extras and deflated for credits.

- A repetitive takeoff is not more accurate than a one-time, carefully structured joint effort (contractor and reviewer).

DEFINITIONS

Wish list or conceptual estimate:
An interpretation of costs from very conceptual information, sketch or program description based on SF, volume or system evaluations.

Preliminary estimate:
A line item estimate based on 50-75% completed design.

Budget estimate:
One or both of the above.

Pre-bid estimate:
The first detailed estimate prepared at the completion of the design, but before contractors/subcontractors get involved.

Bid estimate:
As above, but with contractors'/subcontractors' involvement.

PCO/CO estimate:
An overview estimate or an intrinsic value assessment.

CM baseline estimate:
Same as budget estimate.

CM/PCO estimate:
Same as PCO estimate.

CM pre-construction estimate:
Same as pre-bid estimate.

CM/GMP estimates:
Same as bid estimate, combined with PCO estimate requirements.

An institution should have these estimates follow the CM (construction management) format, i.e. broad-brush, type A, GMP system takeoffs, with fewer line items and system unit pricing adding to a total. In addition, the entire building system should then be shown as a cost per system, single-line item. An example, after HVAC has been broken down into five items:

- compressors   EA
- fan coils   EA or LF
- air handlers   EA
- piping   Total LF
- electric wiring and hookup   <u>LS</u>

                                          $135,000

A summary listing should be provided as follows:

- total cost per SF =

- total cost per ton =

This type of summary item will allow a one-time builder and its CM to judge how a particular design compares to norms. Statistical norms are made available through many sources and publications for quick comparison.

ESTIMATING BY AVERAGES (PART OF BROAD-BRUSH ESTIMATING)

Under this process of broad-brush estimating, two estimators (one senior and one trainee) work as a team on the entire project. The process relies more on statistical data than actual unit pricing. The data reflects the accumulated information by the estimating entity, which shows a percentage of the total cost for systems and subsystems for "typical" projects that can be identified to have similarity with the project being evaluated. This statistical data also maintains realistic numbers for total systems and subsystems. The actual procedure can be described as follows:

1. The more senior member of the team evaluates the general conditions on a preprinted form, based on location, duration, accessibility, supervision and management requirements and other items, normally used to determine indirect costs, not soft costs.

2. The estimator/cost engineer determines what percentage the GC should be (7% to 14%) based on job location, design complexity, accessibility, duration, etc. The total or likely cost is then projected, based on this determination, and that total is noted on the GC sheet for future comparisons as a target final number.

3. The team prepares a sheet together with statistical information on a preprinted form, which contains:

   - constructed area
   - renovated area
   - façade area
   - number of windows
   - number of doors
   - SF of partitions
   - roof area
   - LF of coping

- number of bathrooms
- number of plumbing points
- more basic information

This sheet is used to establish approximate quantities for systems and subsystems to be priced as per historical data accumulated for this purpose. Calculations are done on the spot without worksheets that would require transposition.

4. When the statistics sheet does not provide information for number 3 above, actual quantities are established without worksheets by one estimator measuring and another estimator using the calculator. Again, no worksheets are required and two individuals check the quantities on the spot.

5. The sub-summary sheets thus created are to receive a "use" or judgment number, which can be higher or lower than the actual calculation(s), based on the senior estimator's judgment. The sum of these "use or say" numbers are listed and compared to the anticipated target total, as in number 2 above. The senior estimator adds a contingency, as may be appropriate, based on his "feel" for the final cost.

Please note: This system of costing can be used for all types of comparison estimates or even bidding estimates, but cannot be used for creating a material purchase list or a subcontractor's bid.

— This page for annotations —

VII

SCHEDULE REQUIREMENTS

In order to produce a meaningful cost projection (an estimate), it is best to lay out the work/scope in a logical progression and, thus, create a sequence of events that leads to a critical path schedule. Such integration of scheduling and costing is very important to a project owner/sponsor since both cost and time are an important goal. Further, after a contract has been formulated, an owner/sponsor will require a cost-loaded CPM (critical path method) schedule to track progress and payment requisitions. It is for this reason that all contractors must think ahead and formulate a cost-loaded schedule during the early stages of involvement in the project (the bidding stage). The preferred scheduling software to be provided is the Primavera Project Planner, but any less-complex system will do.

The CPM is to be based on a Precedence Logic Diagram, which sequentially links all activities in the project. All key submittals, approvals, procurements, mobilization, construction and closeout activities must be shown. Schedule activities must show duration, early and late start dates, total float and responsibility. A simple bar chart, in lieu of the above, is acceptable if a Primavera System cannot be used to its full potential.

— This page for annotations —

VIII

INTERACTION WITH A COST TEAM

An owner/sponsor should deal with his contractors on a collaborative level rather than a "them and us" posture. For that reason, and in order to cut costs, he should establish a cost engineering and technical evaluation section (CETES). This section's functions are predominately to help the owner/sponsor expedite the construction program and control costs. The section should also be made available to the owner's/sponsor's contractors to help them better deal with cost projections and constructability issues. The owner's/sponsor's GCs, CMs and subcontractors should be urged to avail themselves of any existing expertise. The following was at one time published as a self-contained primer on the topic:

It is not common for every organization to have a group of cost engineers acting as a support group for the organization. Most have an estimating department, or worse, a few estimators who act as individuals and not as a unit. Less common is also the assumption that a group (estimators or the more advanced levels, cost engineers) can help individuals within a given organization with more than an estimate. By virtue of their vast experience, cost engineers may steer PMs clear of many trouble spots, which include:

- budgeting problems
- scheduling problems

- scope interpretation problems
- contract inclusion problems

This writing was originally prepared to assure that the owner's technical personnel know what has been made available to them, when and what they can ask for, and what they may expect to receive from a unit that is officially known as cost engineering and technical evaluation section (CETES).

Primarily and foremost, such a section should be available to prepare cost evaluations for new work. Secondly, the section can evaluate and suggest alternate methods of construction:

- to save time
- to save money

The section can also assist an in-house PM with structuring a favorable schedule. More importantly, the section can assist others in obtaining work. One of the services cost engineers can provide is ID/C CREA, an integrated service for the design process (previously discussed).

It should also be understood that a cost section/ department is not a dump-all ground. Although PMs may (and should) feel free to call on CETES for technical and cost assistance, he should not call on the section to solicit bids or anything that he can do himself or have non-estimators do. Such a section is staffed with costly help. Use its services, but use it wisely.

A PM or business-development group may approach the section with a clear definition of a problem or need. Preferably, that problem should be summarized in a short statement and with a time frame included (a handwritten note is acceptable). CETES will weigh its workload and offer a delivery date. If acceptable, a written report will be delivered on time with or without a verbal explanation. The information delivery format has been standardized and includes:

- a cover sheet identifying the job by name and number and the person requesting the work

- a clarification sheet showing inclusions, exclusions, exceptions, etc.

- cost reduction suggestions, if applicable

- a summary of costs

- individual trades or items priced

No worksheets are made available. The following paperwork stays in the section for cross-reference:

- a wish list
- a visual conception
- preliminary design
- working drawings

Change orders (COs) can likewise be evaluated from the same criteria. Therefore, PMs can contact CETES before any CO is contemplated.

Further, contractors are to be assured that such a section and its instituted procedures will not introduce another layer and perceived costly consequences.

— This page for annotations —

IX

IS COST REDUCTION ANOTHER LAYER IN THE DESIGN/CONSTRUCTION PROCESS?

In the early stages of planning a project, the management of the owner/sponsor is faced with the reality of choosing a site, an architect and frequently peripheral consultants to assure that their project is successful. Such consultants can be legal, accounting, real estate (rental/sales), landscaping, signage and cost control/cost reduction. Often the process of coordinating all those involved can seem overwhelming. Many owners/sponsors minimize the number of advisors for that very reason. The question then arises: Does the project manager serve his interests well by not seeking all of the expertise available? The other side of the argument may be: Is having too many consultants detrimental to a project's progress?

The following information attempts to dissect this dilemma. It concentrates on assessing the inclusion of engineering/construction cost reduction versus omitting such efforts for the sake of reducing the number of layers created in the management of the design/construction process.

As in most situations in management, too little or too much is a detriment. Too much information can delay decisions. Too little information can result in making the wrong decision. The manager must decide how much is enough. Basic needs are obvious: An architect and mechanical and structural engineers are a necessity. The other consultants may be viewed as a luxury. The manager is faced with assessing which luxury is really a necessity to make his project meet his needs. If legalities are complex, a lawyer is a necessity, not a luxury. If cost and runaway budgets are a concern, a cost reduction/value-engineering team is a must.

A cost reduction firm is attuned to design requirements (the owner's/sponsor's needs, his program), the available funds (the owner's/sponsor's financial position), the life- cycle costing implications (operators ability or desire to incur replacement expenditures) and the design time as a cost factor (delays in designing and construction costs, direct or indirect expenditures). Out of necessity, any cost reduction suggestion must take into consideration the time it will take to implement the suggestion. Redesign costs and drafting, estimating and holding time should likewise be taken into consideration.

An owner/sponsor would never allow his design team the freedom to design anything they like without establishing the need (the program) as a guide. Yet, the same owner/sponsor, in effect, may be doing the same to his budget if he does not avail himself of full cost control through the use of a value engineer. A VE will assure that the architect and engineers use the best system available for the cost range required to ensure that the owner/sponsor gets one hundred cents value for every dollar spent.

If a cost reduction program is to introduce lower costs and improve values, it cannot introduce delays. If it does, or if the manager needs additional time for interacting with a value-engineering team, the VE team is really a cost inducing team, not a cost reduction one. The very existence of a value-engineering firm is to introduce speed, efficiency and economic responsibility. To assume otherwise is like assuming that a medical rehabilitation program creates drug abuse. Cost reduction engineering and creating delays and/or additional layers of administrative responsibility are mutually exclusive.

A cost engineering, cost reduction team must accelerate its own actions to correspond to the activities of the others involved and then exceed the established pace to induce others to fall in line. A good cost engineering team never requests that the design pace be halted so that it can produce evaluations and recommendations on what is available. It must and, of necessity, does work interstitially with the designers as they progress.

The team's recommendation is precise and clearly presented so that the manager does not have to spend time reading and interpreting the intent. Most of the give-and-take required to test cost effective alternatives takes place between individual design team members and the VE team.

The United States Army Corps of Engineers' studies suggest that a twenty-to-one ratio is a statistical probability. That is, for every dollar spent on cost reduction fees, the owner is expected to realize twenty dollars in savings. The

Engineering News Record (ENR), a professional journal, suggests in an article in the March 15, 1990 issue that a forty-to-one ratio is possible. Some VE firms, when dealing with clients, will guarantee a ten-to-one ratio as an absolute minimum. This guarantee does not include an estimated quantitative amount for the professionalism that cost engineering introduces into the design process, nor the benefits derived from the perception others get (investors, approval authorities, board members, endowment groups) that the management entity really cares how the money is being spent and that full value is being sought.

When all pros and cons are carefully weighed, it quickly becomes apparent that the administrative effort required to integrate a cost engineering team is minimal, that a cost engineering team is not a layer in the design/ construction process and that a cost engineering team produces savings equal to a multiple of the fees it receives. No manager can afford to dismiss VE involvement for fear of too much paperwork being generated or for reasons that an extra consultant will be a drain on his time. Cost engineers (CCRE, certified cost reduction engineer) relieve the manager of many responsibilities, which they can do quicker and more effortlessly. A good CCRE team works in the background and only generates work for themselves. The results are presented only to management, who is left with the (yes or no) decision-making process. The management is not required to spend time doing legwork or extensive research. All cost related facts are laid out in front of them.

Universal cost reduction and cost control procedures are here to stay. Contractors are urged to work within the system rather than fight it. An owner should enlist all the positives of the various procedures available and urge his contractors to do likewise.

— This page for annotations —

X

THE PYSCHOLOGY OF NUMBERS USED IN CONSTRUCTION

The psychology of numbers in construction is a tool that many contractors rely on, that many designers fail to anticipate its effect, and many owners/users can utilize to their advantage. How a project is perceived affects the cost of the work and, by extension, its feasibility. To obtain an economic platform from which bids can be negotiated is as important as a good cost engineered design. Builders/owners can create such a platform by adhering to three principles:

- The pre-construction services (PC/S) should be performed in-house or, at least, under in-house direction. To assure that contractors perceive a project as a viable economic structure, they must be convinced that the owner knows its intrinsic worth, that he understands the constructability implications and that he has a good feel for the prevailing market conditions. Giving a contractor/CM an outer-limit estimate, prepared as part of the PC process, establishes such a perception. This outer-limit estimate should be coupled with verbal orientations to reinforce the owner's control and demonstrate his involvement.

- An owner/user must never allow a potential GC/CM to assume that he is the only viable contractor. This could allow him to give notice directly or indirectly to everyone to stay away.

- An owner/user must never allow the design team to deal directly with the potential CMs or subcontractors for the purpose of sourcing special materials and establishing costs. Designers speak a different language and this approach fragments each system and makes negotiating a complete package difficult and, hence, an economical result unattainable. To price subcomponents from several sources increases the cost and coordination, and fragments the responsibility issue.

The total psychological setup can be responsible for as much as 30% of the total projected cost. "Good" psychology means a lower-base estimate, and "bad" psychology means higher-base proposals, which act as the jump-off point for further negotiation(s). The very basic argument for establishing a psychological edge starts with demonstrating an owner's control, his determination not to let contractors/ subcontractors take advantage of him, his ability to "feel" market conditions and his ability to "speak" the contractor's language. The total process can be accomplished with in-house personnel, if it is available, or outside help, which must act as if they were part of the owner's staff. Contractors can be directed to be realistic, but primarily they must be convinced that an owner knows what he wants and needs.

— This page for annotations —

XI

COST ENGINEERING AND RELATED IN-HOUSE REQUIREMENTS

This chapter is mainly for in-house use. It should be used by an owner's/sponsor's PMs for guidance and awareness. It may be pointed out to a contractor as a reference document or it may be ignored. Since an owner should strive for a collaborative effort with all its contractors, this information may be used to demonstrate that all involved in the construction process understand the underlining philosophy.

This chapter's information is a result of an investigation into the existing cost control practices, or lack thereof, its shortcomings and/or improvement potential from the prospective of cost engineering, estimating and integrated cost reduction engineering (ID/C CREA).

Although most projects handled by institutions/ developers are simple, some complex projects do exist (examples: museums, housing complexes, colleges). The requirement/format for cost control/cost engineering is not the same for all types of construction undertakings. There is always enough commonality in all types of construction to warrant creating a process, which can serve both major and less-complex undertakings with minor variations. The thrust of what may be needed to improve the overall monetary overview should be:

- estimating/cost evaluation
- integrated design/cost reduction (ID/C CREA)
- constructability and completion (schedule) controls

To gain a strong handle on costs and time, which translates into cost, some estimating projections are required. This can be accomplished with a PC/S department with a minimal workforce and a low-operating budget. Since a PC/S contract with a CM may already be in place, only an overview of this process need be instituted. For future projects, large or small, an owner should form his own department with the help of a part-time consultant or outsource such functions to a practitioner to provide all PC/S.

Perception in the (New York) construction market is as important as actual working systems in place. How contractors perceive the owner/sponsor is handling and controlling his actions will determine how his intentions are executed. Therefore, it is important to create the proper climate for cost efficiencies. The climate must include a demonstrative posture that a given owner has the resources to do his own pre-construction services, the ability to evaluate subcontractor's quotes and the ability to evaluate change orders. A pre-bid meeting, where the owner can discuss and emphasize his cost concerns, is an important ingredient in building the right perception and should be instituted by all PMs. (See Chapter II.)

Based on observations within this segment of the construction industry, review of many documents and the general climate, it is suggested that an owner should institute the following:

- Expand his information system to include reports for:

  - total hard cost
  - total in-house costs (PM, clerical)
  - total soft cost
  - total final/SF cost
  - total overages (costs above estimate)
  - total under estimates
  - time projections (over and under)

- Create a reading file into which all GC and CM correspondence, to and from subcontractors and agencies, is placed for periodic review by the in-house PM. This will alert an owner of pending costs or other problems.

- Set all schedules on a common footing (preferably the Primavera Project Planner system) that would allow the combining of cost and time (a cost-loaded schedule). Note: Historically, this requirement cannot be forced on small GCs due to their limited ability to use complex schedules such as Primavera.

- A bid form with an in-house takeoff should be given to potential GCs or CMs and their key subcontractors at the pre-bid meeting. The meeting should be set up for each project in order to create a climate for cost concerns, cost reduction, alternates and bid-form adjustments, as well as creating a favorable perception of the owner's abilities. A good psychological negotiating weapon with a GC is a very professional-looking takeoff summary in report form with all times listed. The thicker the report is, the more potent the weapon.

- Before a preliminary design is handed to a GC for a proposal, a wish list in-house budget could guide the GC in his cost evaluation. It is highly suggested that this type of budget can be created from very basic facts with the help of the cost evaluation section.

- Implement an ID/C CREA program, which may include the designated PM and GC, the in-house PM and the designated outside or part-time cost engineering entity. ID/C CREA is an instantaneous cost evaluation assistance program offered to the design team on an ongoing basis. For further clarification, see Chapters IV and V.

- A recommended (in-house or out-sourced) department, to be named cost engineering and technical evaluation section (CETES) incorporating a full-time estimator and partial use of a PM, draftsmen and purchasing agents, should be established and assigned to any new venture. Further, to make this department function properly, an outside cost engineer or a cost engineering entity should be utilized on an "as need" or retainer basis to act as a coach and trainer and to be the catalyst to make the functions run smoothly. The estimators could be trained to do complex takeoffs. The senior cost engineer could then create a report to be used, as described heretofore, including change order evaluations and requisition reviews. This department could also hone the (current) bid form(s) and offer the guidance required to obtain favorable bids from GCs and subcontractors. The part-time outside cost engineer could also perform all pre-construction services and train the full-time estimator(s), project manager and CM in the art of understanding the value of PC/S. To allow the CM to do pre-construction services is to forfeit valuable opposing views and the main thrust of the work, the owner's interest. This format for a partial department is being suggested because:

- Its formation would create minimal corporate outlay.

- It would utilize the spare time of the existing technical staff.

- It could obtain higher-caliber help than would otherwise be available for the equivalent cost.

- It would induce the potential GC/CM to take on an opposing view with regard to constructability, costing and subcontractor negotiations, and cost reduction. Under most current PC/S contracts, the CM need not do what may be good for the owner/sponsor arrangement; he may guide the design and costing toward what is only good for him.

- The mere existence of a cost section could act as a psychological exclamation point, highlighting to potential GCs and CMs that an owner intends to, and is able to, look into all costs with more than superficial vigilance.

— This page for annotations —

XII

CHANGE ORDERS: PROCEDURES FOR PROCESSING CHANGE ORDERS DEFINED

INTRODUCTION

The project managers on any major project are the front-line leaders, who make a project progress, who make the work happen, who enforce the schedule and who carry the ultimate responsibility of delivering a completed project. PMs become PMs because of their vast experience, their management abilities and their organizational skills. To help them achieve their goals, PMs need a support staff, good channels of communication and, above all, procedural guidelines that make all actions uniform, easy to follow by the support staff and conform to the needs and wishes of the client. When a project has the total structure divided into sections, where several PMs have to work in close cooperation and where changes are numerous, the need for uniformity becomes even more important and highly desirable. Toward that goal, this chapter outlines all procedures as they relate to standard PM activities and change order processing, in order to present a professional posture and to process quickly and painlessly all PCOs (potential change orders).

PROJECT MANAGEMENT/ CHANGE ORDER PROCESSING

How to manage a project is a topic on which much has been published. University courses and degrees in construction management have been available for more than forty years. The "how to," therefore, is beyond the scope of this writing. It is further taken as a given that PMs, when they are assigned to that status, are at the top of their profession and do not require being told how to do their work. What this writing is meant to impart is a standard process to help the paper management, especially as it relates to PCOs. Because of the numerous changes, clarifications and potential changes that occur on any major project, many PCOs accumulate. To avoid a clerical and administrative deluge, each PM must use all available channels and methods that have been established to help unclog the administrative arteries. Above all, the PM must be aware that:

- Only the PM is authorized to negotiate a CO for extras or credits.

- Only the PM should keep the subcontractor "honest" and within acceptable bounds of CO pricing.

- Only the PM can threaten a subcontractor with contract cancellation.

- Only the PM, with the estimating department's help, can prepare a formal change order proposal to be submitted to the owner or program manager. Failure to do so may cause his employer to lose funds to which the employer is entitled.

- Only the PM will have to answer to the owner's representative if the formal proposal is not complete or accurate.

- Only the PM represents the contractor/CM. How he behaves reflects on the company he represents. Therefore, incomplete PCOs for reason of inaccuracy, for improper evaluation or for being too lenient with subcontractors, as a whole, do not reflect well on his employer.

PCO CREATION AND PROCESSING

Change order requests (CORs) and potential change orders (PCOs) are created in the following sequence:

- PMs are to look out for work which may surface that is not contract work, i.e.:

  - new or altered design(s)
  - contractor initiated omissions
  - designer initiated changes
  - CM initiated convenience changes

These changes may constitute a potential extra or credit.

- A PM, who wishes to register a PCO, must complete the form provided for that purpose.

- The PCO is identified and subsequently evaluated. The PM obtains a subcontractor's proposal for the work involved.

- CM/GC evaluation process calls for the PM, with his estimating department's assistance, to correlate the cost with the subcontractor(s), negotiate such cost or credit and obtain a revised proposal. The PM must not reveal to the subcontractor his assessed value, as it will dilute his ability to negotiate a favorable price.

- After all backups has been honed, revised and updated, a contractor's statement/proposal is prepared on forms provided.

- The proposal contains individual values that were hammered out and honed by the PM's estimating department.

Although others may sometimes prepare proposals, they are the responsibility of the PM and should be reviewed and approved by him to assure that all facts are correct, that the evaluation is reasonable and that his employer's interests have been fully protected. For that reason, every estimating summary contains a reference to the date when the PM was requested to give his concurrence.

As stated, any PCO must start with the PM who can discourage the originator (the designer or subcontractor) from submitting it and making that request official. If the PM deems a particular change is warranted, the estimating department, through the PCO record-keeper, will require the PCO requisition form (a simple information sheet, see Chapter XIV) and, as a result of this formal request, will assign a PCO number and enter it into the log. This automatically will put the PCO in line for evaluation and review, with the PM to set strategy for the continuation of the processing procedure.

It is very essential that the logged PCOs be processed quickly. Accordingly, the estimating department will limit its turnaround time to no more than a bare minimum and, upon return to the PM, it should not be in his possession for more than five days. Upon a quick review with estimating, the PM will request that estimating prepare a draft PCO on the forms provided. Estimating will review the PCO within twenty-four hours and return same with:

- clarifications and/or
- correction (of facts) and/or
- recommendations and/or
- a revised estimate and/or
- a revised, corrected contractor's proposal

The PM is then expected to review the subcontractor's proposal(s) for the work involved and to submit a final version of the change order, review this final version with estimating and submit the final document so that estimating can present it to the project director for signature and final submittal. At this stage, the project director should not have to review numbers, factors or ascertain completeness, but only sign the document. Estimating will then submit by hand the signed document to management for further processing. In case of a bulletin or a multi-contractor PCO, the PM affected by the largest dollar change becomes the lead PM responsible for the submission process, as if it involved only his portion/area of work.

Estimating, which encompasses the keeping and updating of the PCO log, should exist for the sole purpose of assisting the PM's functions, a portion of which is evaluation of changes and requesting change orders for work not considered part of the original scope. Thus, all PMs are urged to avail themselves of the services that include the assessment of non-PCO changes, such as cost reduction suggestions (VE-CREA) and pre-buyout evaluations. Estimating should issue bi-monthly log updates to each PM, who is requested to review his subcontractor's outstanding COs and inform the log keeper of any inaccuracies. The PM should use the log as a reminder of what is outstanding and what needs "clearing the deck."

Final comment: There is no reason for PMs to accumulate subcontractor claims without processing them through the established channels and forwarding those claims to the owner, if they are legitimate. Estimating exists for the sole purpose of assisting the PMs to process paperwork. There is no more important paperwork than processing COs to facilitate requisitioning money due. Put differently, PCOs are important for job progress, cash flow and keeping subcontractors current.

In processing CORs (change order requests), it is important to indicate proper reference. The important designations are:

- PCO number
- CO number
- bulletin number

This somewhat complex procedure was utilized on Terminal IV at J.F.K. International Airport. Smaller projects do not require such formality. The process, although less complex, still needs to follow the same steps: PM to estimating to subcontractors and back to PM for a final COR from the owner.

— This page for annotations —

XIII

AN ESTIMATING DEPARTMENT PROCEDURAL MANUAL

A PROCEDURAL OUTLINE: HOW THE ESTIMATING DEPARTMENT OPERATES

I.   The way estimates/value assessments are generated, quantities are obtained and CM estimates are used, is outlined in various writings by this writer to be used as an information or training tool.

II.   Major estimating departments are functioning as a barometer for testing contractors' proposals and/or change orders (omissions and extras) by providing intrinsic worth assessments. Estimating is not empowered to negotiate contracts, but to develop intrinsic values of work to be negotiated by others. COs or scope changes are also its main function, which encompasses the aforementioned. It also exists to evaluate PCOs and assist PMs in resolving owner/designer generated changes (additions, subtractions or scope changes) and help process CORs through the maze of requirements.

III.   Estimating departments usually monitor a PCO log to track changes to the various costs and to establish workload requirements.

Estimates are provided:

- When required for COR submissions or when the change is valued at $50,000.00 or more, as may be appropriate for a given project.

- When a PM is not at ease with what the contractor offers and requests or requires a yardstick of values for comparison.

Priorities for providing assessed evaluations are established on the following basis:

- The chronology of a bulletin, change request or clarification, as it was generated.

- The size/amount of the PCO or COR.

- The PMs need to resolve lingering issues.

- The time period the PCO has been under review.

- Any urgent "resurfacing" of a PCO issue by new events or developments.

IV. The chief estimator's duties are to sort out the PCO log, the PMs requests and the project director's wishes, and assign PCO priorities and thus establish the department's workload. He is also the head coach who trains and hones the individuals to act as a functioning unit. He performs actual work when conditions warrant. The chief estimator also establishes policy on what constitutes a "fair but reasonable value" by guiding the evaluation process. He is also responsible for the final submission of the department's work. Estimates are prepared by individuals and during "round table" sessions, which are scheduled two to three times weekly. During these sessions, an estimate is handled as a joint effort with all estimators participating in interpreting the change, in quantity surveying, in pricing and in summarizing a final statement. This statement is prepared in two forms:

- A simple form, which consists of one or more worksheets or sub-summary sheets and a summary sheet, listing an owner's assessed amount and comparison to the contractor's evaluation.

- If the PCO is complex, has an old or long history, differs greatly from the contractor's evaluation or is a candidate for protracted negotiations, the estimate will have all of the above, plus a background, history and the divergence of the values reached.

V. When an evaluation has been completed, the results are placed in the PCO folder and in the estimating PCO folder file.

Two reports are issued periodically:

- a PCO estimating status.

- a PCO log information sheet (see Form 005, Chapter XIV), giving each PM effected a notice that they are holding a given PCO, which needs processing and COR submittal.

VI. A new-type PCO log now exists. It lists PCOs that require an estimate for COR submission. It is often being used to spread the estimating workload. Not shown on it are PCO assessments, which are valued at below $50,000.00 or which a PM or the project director wishes to evaluate to keep the contractor "honest." In addition to the PCOs listed, estimating will prepare all evaluations that may be required by any member.

VII. PMs are to be reminded constantly that, in addition to estimates and value assessments, the estimating department is ready to assist with COR procedure and VE/CREA (Value Engineering/Cost Reduction Engineering Analysis) evaluations. With the latter, a PM may obtain any suggestion for cost savings from any source and have that idea translated into dollars now and over the life of the project. Also, any PM may request estimating to analyze any segment or system and come up with CREA suggestions in order to have these suggestions priced and presented to the owner for consideration.

VIII. After a PCO has been submitted and rejected, the estimating department checks the owner's comments to ascertain that an evaluation is (or is not) required and complies with those requirements. Thereafter, the estimating department follows up with the PM involved to assure that the PCO in question gets resubmitted and/or resolved. This "bird dog" duty is one that is placed with the estimating department by the project director. Accordingly, once each week the rejected/returned PCOs are being distributed to all estimators for the purpose of follow up with the project managers.

IX. How to file and process a PCO has been outlined by way of pre-prepared forms. It is based on the estimating department's premise that PMs must involve the contractor in any COR process by negotiating with him to obtain a fair proposal for the work in question in the amount the owner can live with.

— This page for annotations —

XIV

ATTACHMENTS:
STANDARDIZED FORMS,
REFERENCES AND
SELECTED READING MATERIAL

Contents

WISH LIST QUESTIONNAIRE

The following information should be assembled for the preparation of a realistic intrinsic worth for a contemplated design/construction undertaking.

Who is the likely project manager? _____

Project name _____

Project location _____

Anticipated project size (S.F.) _____

Probable design team _____

Are there any GCs or CMs as likely candidates?   Yes _____ No _____

List three _____ _____ _____

Can they be contacted now?   Yes _____ No _____

Is there any money already allocated?   Yes _____ No _____

Is there a very specific need (describe)?
- Architectural statement? _____
- Material statement? _____
- Special flooring?_____
- Special wall covering? _____
- Special store hardware/fixtures? _____
- Special acoustic requirements? _____
- Special vibration requirements?_____
- Other _____

Is there a similar project that will be used as a model?

Name   _____

Location   _____

Manager   _____

Phone #   _____

A budget is required by (date) _____

Format (please check one)
- Formal, standard format   _____
- Informal, in memo form (totals only)   _____
- A verbal estimate   _____

ESTIMATING FORM

This form is to be used as a takeoff sheet when providing a Type A, GMP estimate or as a summary sheet when individual quantities are to be created (refer to CETES Manual, Vol. B).

Project: _____ Total project S.F. [　　　　]

Location: _____

System identified: _____

Trade, Ref. or CSI #: _____

Date: _____

Notes and/or special comments: _____

_____

_____

Ref/CSI #	System/ Subsystem Description	Quantity	Unit	@	Cost

Actual total [　　　　　　]

Estimator's suggested amount to be used [　　　　　　]

Cost Engineering and Technical Evaluation Section   Form 001a

ESTIMATING FORM SUMMARY

This form is to be used as a takeoff summary sheet when there is more than one estimating sheet (Form 001).

Project: _____ Total project S.F. ☐

Location: _____

System identified: _____

Trade, Ref. or CSI #: _____

Date: _____

Notes and/or special comments: _____

_____

_____

Ref/CSI #	System/ Subsystem Description	Estimator's Suggested Cost	Comments

Estimated project total cost per system ☐

Estimated project per S.F. cost for system ☐

Market norm per S.F. cost for system ☐

GENERAL CONDITIONS MATRIX

ITEM	BY ONTRACTOR	BY TRADES	BY OWNER
Principal-in-Charge			
Construction Executive			
General Superintendent			
Project Manager			
Project Superintendent			
Area Superintendents			
Architectural Field Superintendent			
MEP Field Superintendent			
Mechanical & Electrical Coordinator			
General Labor/ No. ( ) of Full-Time Laborers			
Clerk of the Works/Timekeeper			
Shopkeepers			
Safety Supervisor			
Project Engineer			
Project Accountant			
Secretary			
Accounting Support			
Warranties			
Scheduling			
As-Built Drawings/Closeout Documents			
Cost Engineering Assistance to Owner			
Estimating for Bidding Purposes			
PC/S Assistance to Owner			
Trash Containers/Dumpsters			
Final Broom Clean			
Rubbish Carting			
Off-Hours Carting & Deliveries			

ITEM	BY ONTRACTOR	BY TRADES	BY OWNER
Hard Hats/Rain Gear/Field Supplies, etc.			
Sidewalk Bridge/Hoist/Service Car			
Crane/Scaffolding			
Safety Rails/Barricades & Protection			
Fencing			
Local Trucking			
Temporary Roads & Sidewalks			
Temporary Protection & Barricades			
Dewatering			
Testing & Inspections			
Licensed Surveyors			
Permits & Expediting			
Controlled Inspection & Outside Consult's			
Weather & Winter Protection			
Dust Control			
Temporary Heat/Light/Power/ Water			
Field Office (Inside Building)			
Site Office Expenses/Materials/ Supplies			
Trailer			
Temporary Toilets			
Exterminator			
General Office Supplies/ Operations			
Copy Machine			
Site Telephone & Fax			
Two-Way Radios			
Postage & Express			
PCs/Software			

ITEM	BY ONTRACTOR	BY TRADES	BY OWNER
Progress Photos/Cameraman			
Temporary Job Signs			
Personnel Expenses			
Petty Cash/Local Travel Expense (for GC personnel)			
Guard Service			
Fire Extinguishers			
First Aid Supplies			
Shanties/Storage Equipment			
Rental Equipment/Tool Depreciation			
Non-Expendable Tools & Equipment			
Small Tools & Expendables			
Operating Engineers (Material)			
Elevator Operator (Personnel)			
Teamster Steward			
Taxes			
TOTAL GENERAL CONDITIONS			

INSTRUCTIONS FOR SUBMITTAL OF ESTIMATES

The estimating department will provide and/or review all estimates in connection with bids or PCOs received from contractors for accuracy and/or reasonableness. To avoid delays in processing, the estimate preparation must follow a unified format:

- Identify project, specific location, specific condition(s) and reason for submittal.

- List basic items/systems only with unit costs, including L&M (labor and material), but no OH&P (overhead and profit) or general conditions.

- Provide a line item for discounts or credits, if any.

- Provide a line item for OH&P, as per pre-agreed allowable percentage.

- Provide a net total, add or deduct. (See forms 001 and 001A.)

CREA COST STUDY

This form is to be used for preliminary analysis of CREA suggestions. It should be used to create first-impression values for VE suggestions and used in conjunction with Form 001.

Project: _____ Total project S.F. [   ]
Location: _____
System identified: _____
Trade, Ref. or CSI #: _____
Date: _____
Notes and/or comments: _____

_____

_____

DESCRIPTION	UNIT	@	COST
Substitution Suggestion			
Intent/Original System			
Gain/Loss			
Substitution Suggestion			
Intent/Original System			
Gain/Loss			
Substitution Suggestion			
Intent/Original System			
Gain/Loss			
Substitution Suggestion			
Intent/Original System			
Gain/Loss			
Substitution Suggestion			
Intent/Original System			
Gain/Loss			
Substitution Suggestion			
Intent/Original System			
Gain/Loss			

Cost Engineering and Technical Evaluation Section   Form 003

LIFE-CYCLE STUDY

This form is to be used in conjunction with Form 002.

Project: _____ Total project S.F. [ ]

Location: _____

System identified: _____

Trade, Ref. or CSI #: _____

Date: _____

Notes and/or special comments: _____

_____

_____

Suggested Substitution	Life Expectancy	Est. Maint. Cost Per Year	Initial Cost (From Form 002)	Total Life-Cycle Cost

A. Life-cycle cost for alternate system/subsystem [ ]

Original System/ Substitution	Life Expectancy	Est. Maint. Cost Per Year	Initial Cost (From Form 002)	Total Life-Cycle Cost

B. Life-cycle cost for original system/subsystem [ ]

C. Life-cycle gain/loss alternate system/subsystem [ ]

Acceptance of value-engineering alternate

_____

Signature   Date

COST ASSESSMENT AND SUMMARY FOR PCOs

(SEE WORKSHEETS ATTACHED)

PCO #: _____

Documentation used: _____

Contractor's/CM's estimate*: _____

Subcontractor: _____
Trade code: _____
Subcontractor's estimate:** _____

Net difference: _____

Contractor's proposal date: _____

Estimate reconciled with project manager: Yes ☐   No ☐

Estimate reconciled with contractors: Yes ☐   No ☐

Telephone call on: _____   Spoke to: _____

Telephone call on: _____   Spoke to: _____

Telephone call on: _____   Spoke to: _____

Remarks:

Additional statistics:

_____
_____

Background and historical facts:

_____
_____

Footnotes:

* Backup, quantity surveys and calculations are on file in the estimating department or attached hereto.
** Proposal and contractor's backup in the estimating department's file.

PCO ESTIMATING STATUS REPORT

The following PCOs have been reviewed and/or assessed and channeled to the appropriate project manager, during the period of _____ .

PCO Nos.   ——   ——   ——   ——   ——
           ——   ——   ——   ——   ——
           ——   ——   ——   ——   ——
           ——   ——   ——   ——   ——
           ——   ——   ——   ——   ——
           ——   ——   ——   ——   ——

This informal pre-estimating search for information is being conducted for record-keeping purposes only. Results will be submitted periodically to:

- All estimators
- Keeper of the records
- Estimating department workload schedule
- All project managers involved for further processing

Further, this status report is being submitted for the purpose of adjusting the PCO log and establishing and distributing the workload for the estimating department.

PCO LOG UPDATE INFORMATION

I.   The following PCOs are listed on the PCO log as requiring the CM's review and/or submittal. During the normal process of estimating the workload assignment, it was noted that many of these folders:

- are located with the project manager(s) involved or
- do not require estimates/assessments.

II.   Project managers keeping these folders are urged to:

- process a change order request or
- see the estimating department, which can assist in the PCO's final resolution and disposal.

III.   Project manager(s) and PCOs involved

_____	_____	_____	_____	_____
_____	_____	_____	_____	_____
_____	_____	_____	_____	_____
_____	_____	_____	_____	_____

This informal pre-estimating search for information is being conducted for record-keeping purposes only. Results will be submitted periodically to:

- All estimators
- Keeper of the records
- Estimating department workload schedule
- All project managers involved for further processing

Further, this status report is being submitted for the purpose of adjusting the PCO log and establishing and distributing the workload for the estimating department.

MEMO *(to be issued for reinforcing CO procedures)*

To: All PMs/APMs
From: Upper Management
cc: Keeper of the Records
Date:
Re: • PCO Initiation
      • Negotiation Documentation
      • Preparing, Configuring and Routing
        COR Packages

Utilize the attached forms and form letter to initiate PCOs and to prepare and route CORs for signature, per the following procedures:

I.   <u>PCO Initiation</u>

Most PCOs are initiated upon request from subcontractors, the owner, RFI (request for information) generated changes, etc. However, if a PM/APM wishes to initiate a PCO to document a forthcoming potential change, he can do so by completing the attached PCO Requisition Form and submitting it to record keeping. In turn, the record-keeper assigns a PCO number to the request and enters the information from the form into the PCO log. He will then complete the action block at the bottom of the form, enter the PCO number at the top (if appropriate) and return a copy of the form to the requester.

II.   <u>Negotiation Documentation</u>
        Utilize the attached Negotiation Record Sheet Form
        to prepare for, and document, the results of change
        order negotiations (including credits).

III.   <u>Preparing, Configuring and Routing of COR
        Package</u>
        1.   Complete the attached Contractor's Statement
                and Contractor's Statement Checklist forms.
        2.   Using the attached format, prepare the COR
                cover letter.
        3.   Prepare a signature package with the contents
                sequenced in the following order (from top to
                bottom):
                •   Contractor's Statement
                •   COR cover letter
                •   Contractor's Statement Checklist (with
                        backup)
                •   Negotiation Record Sheet
        4.   Obtain the CM's signatures on the Contractor's
                Statement page and the COR cover letter.
        5.   Reconfigure the above package sequence by
                swapping the order of the COR cover letter
                and Contractor's Statement, so that the COR
                cover letter is on top.
        6.   Immediately prior to sending the COR cover
                letter for approval/signature, the record-
                keeper should update the COR log.

PCO REQUISITION FORM

Project: _____   Date: _____

Project Area Affected: _____

Description of Change: _____

Reference Number: _____
RFI Number:
Bulletin Number: _____
Other Number:
Proposal Type _____
(Fixed Price or T&M):
Status: _____

Subcontractor(s) Affected: _____

_____

CM's Estimate: Required _____ **Not Required _____

Name of person requesting PCO Number: _____

Signature: _____

** If under $_____ (a predetermined amount)

(Block to be completed by Estimating)

( ) Request entered into PCO log on _____ as PCO # _____

( ) Requested change will be processed with existing PCO # _____

( ) Request returned and not entered into PCO log because _____

_____ Cost Engineer: _____

CONTRACTOR'S STATEMENT ENUMERATION

COST ENGINEERING AND TECHNICAL EVALUATION SECTION

TO:		PROJECT:		
		PCO NO.:		
ATTN:		CONTRACTOR STATEMENT NO.:		
ITEMS	**DESCRIPTION**		**TRADE CODE**	**COST BEING SUBMITTED**
Project area defined	xxxxxxxxx			
Scope of change	Provide material, labor, equipment and supervision and to furnish and install changes as indicated in xxxxx			
Reasons for change	Supplemental information to the contract documents in design submittals above			
CM's estimate	The CM estimate for the trade(s) is $____			
Sub(s) proposal(s)	Subcontractor(s) cost proposal(s) for the affected trades is $____ (Refer to Attachment A)			
Price/rate analysis	See Estimate Summary Sheet			
Subcontractors affected and trade/ Subcontractor distribution	TRADE SUBCONTRACTOR $____			
				$____
				$____
				$____
	Subtotal			$____
Cross reference(s)	PCO No. ____ Contract No.____			
Time delay, if any				
CM fee	LS			
	Total			$____
APPROVED BY FDI:		**BY: Estimating/CETES Department**		
TITLE:		**BY: Project Manager**		
DATE:		**BY: Project Director**		
		BY: Principal-in-Charge		

(Suggested letter format for CO request submittal)

Date: _____

Attn: Program Manager

Re: Contractor's Statement

Dear _____ :

We are enclosing for your approval Contractor's Statement No. _____, dated _____. This Contractor's Statement is based on the following:

- The scope of work included in this proposed change is beyond the contract scope and constitutes a valid change.
- Time delay analysis indicates no extension of time.

Negotiations with the subcontractor have resulted in a reduced agreed price of $_____.

We have reviewed subcontractor(s') proposal(s) in terms of scope, quantities and pricing, and are submitting the attached statement with recommendations for its approval.

Please issue the appropriate change order so we can process this work for payment.

Sincerely,

Project Director

SUBCONTRACTOR NEGOTIATION RECORD SHEET FOR CORs

Project: _____   Date: _____

Project Area: _____

PCO Number: _____

CSI Number: _____

Estimate reviewed with subcontractor and PM: YES ☐   NO ☐

Negotiation Record:

Item	CM's Estimate	Sub's Proposal	Negotiated Revisions		
			No. 1	No. 2	No. 3
Date					
Direct Costs					
Overhead & Profit					
Subtotal					
General Conditions & Bond					
Total					
Remarks:					

Negotiated Record:

Sub's Initial Proposal $_____

Sub's Final Price $_____

CM's Estimate $_____

Differential (final) $_____

Final price negotiated with sub's representative: _____ Dated: _____

Owner's/Rep Name: _____ Initials: _____ Dated: _____

CONFIDENTIAL

COST COMPARISON EVALUATION

      This form is to be used as an overall summary of cost evaluation applications. It is meant to convey in an abbreviated form how the owner's contractors evaluated the cost with the help of the owner's input. It should be used as a tool for contract award process and be utilized for in-house use only.

Project: _____   Total project S.F. [   ]
Location: _____
System identified: _____
Trade, Ref. or CSI#: _____
Date: _____
Notes and/or special comments: _____

_____

_____

Ref/ CSI #	Item Description	Owner Budget	CM/ Contractor Budget	Over the Norm	Overall Evaluation (Yes/No)

Narrative: _____

_____

_____

_____

_____

CONTRACTOR'S STATEMENT CHECKLIST FOR FORMAL PCO PROCESSING

Area: _____

ESTIMATING/CETES DEPARTMENT   PCO No.: _____ Date: _____

Requirements	Included			Remarks
	Yes	**No**	**N/A**	
I . Cover Letter Includes:				
Change validity statement				
Cost impact				
Schedule impact				
II. C.M. Sheet Includes:				
Work area defined				
Scope of work defined				
Reason for change identified				
Subcontractors identified				
Cost impact stated				
Schedule impact stated				
Cross-references identified				
Trade code identified				
CM's fee included				
III. Attached Backup				
This sheet signed				
Schedule Report				
Narrative				
Subcontractors' Proposal				
Estimate				
Copy of COR				
Copy of RFI				
Bulletin cover sheet				
Sketch/drawings				
Correspondence				
Lump Sum Details				
Contract unit rates				
Labor rates				
Equipment rates				
Material costs				

T & M Details				
Employee name & trade				
Description of work				
Signed by CM's supervisor				
Labor & equipment rates				
Material invoice				
	PM/APM name:		_____	
	PM/APM signature:		_____	
	Date:		_____	

MEMO *(for revising estimated costs)*

To:
From: Estimating Department
Date:

Re: Adjusted Estimates Submitted Herewith

Theabovereferencedestimates/PCOshavebeenreevaluated and are being submitted for your use and information. The reevaluation was based on the following:

I.   The Department's directives to the estimators have been redefined, honed, and adjusted, since these PCOs were evaluated:

- Estimating valuations represent the intrinsic value of the work in question, to be used as a basis for further negotiation with contractors. They are not prepared to fit in a certain position between contractor's estimate and owner's representative (guesstimate). Nor, are we trying to take advantage of any contractor.

- Unit pricing used is now always as per contract, no matter what disclaimers the contractor may offer.

- PCO evaluations are now being prepared to prevent contractors from using PCOs as a vehicle for enrichment. "Clarification" information is considered as part of contract obligations. Excessive unit pricing by the contractor is not allowed to influence the estimating department's outcome.

- Additional general conditions are not allocated for changes valued at less than $\underline{\$}$.

- Costs for drafting, coordination, and special handling are likewise disallowed. The original contract does include overhead, profit, and a certain amount implied for additional coordination and handling.

- In response to the owner's wishes, a policy of being somewhat more aggressive in protecting the CM's <u>expected posture</u>, as it relates to the owner's interests, is expected. In reality, project managers are urged not to discourage contractors from submitting COR's, which are reasonable, but to reserve the right to do so after the project has progressed. Thereafter, "minor" and "insignificant" extras, in terms of the percentage of the contract they represent, should be discourage or disallowed.

II. Contracts should be carefully scrutinized for inclusions, which contractors often ignore. Forms being used oblige each estimator to obtain more clarification from contracts and be aware of lump sum miscellaneous inclusions for items or details they should have known about. "Clarifications," if a change is so judged by the available evidence, is disallowed as an extra. Evaluations worth nil, in our opinion, are being recommended to the project manager for dismissal or disallowance.

III. Since project managers are <u>expected</u> to be more aggressive in protecting the owner's interests and, since the general perception is that project managers need strong backup to deal with subcontractors, estimates should reflect a more caustic posture and a sharper sword for the project managers to carry into battle.

IV. The estimating department is only a provider of fair costs. It is not a provider of a "death blow" instrument meant to harm subcontractors to the point of hurting or hindering overall progress. Still, the project manager is expected to obtain the lowest possible price from each subcontractor before that price is submitted to the owner for approval. Later in the project's progress, the lowest possible price for extras and highest possible amount for credits should be sought, after the owner's change order has been obtained. Estimates may be used as a basis for negotiation in either case.

CONTRACTOR'S STATEMENT:

A CHANGE ORDER REQUEST

CSI No.: ＿＿＿＿＿＿＿

PCO No.: ＿＿＿＿＿＿＿   (Bulletin ＿＿＿＿＿)

CO No.: ＿＿＿＿＿＿＿

Area Affected: ＿＿＿＿＿＿＿＿＿＿＿＿＿＿＿

Total Amount Requested: $＿＿＿＿＿＿

Total Schedule Change: ＿＿＿＿＿＿＿＿＿＿＿

Date: ＿＿＿＿＿＿＿

CONFIDENTIALITY AGREEMENT TEMPLATE

AGREEMENT by and between _____, Owner/Sponsor and the under-signed company (the "Company").

WHEREAS, the Owner/Sponsor and the Company desire for the Company to act as a supplier and/or render services to the Owner/Sponsor in connection with _____; and

WHEREAS, in connection therewith, the Company has been given and may be given access to, generate, or otherwise come in contact with certain Confidential Information of the Owner/Sponsor and/or the customers of the Owner/Sponsor; and

WHEREAS, the Company and the Owner/Sponsor desire to prevent dissemination or misuse of such information;

NOW, THEREFORE, the parties mutually agree as follows:

1. **<u>Confidential Information</u>**. The Company acknowledges that a relationship of trust and confidence has been created between it and the Owner/Sponsor. All documents, material or information (whether written, oral or other form and whether or not displaying proprietary or restrictive markings) which has previously been received by the Company or may in the future be received by the Company, pertaining directly or indirectly to the business of the Owner/Sponsor and its customers, whether received or developed by the Owner/Sponsor, including, but not limited to, the Owner's/Sponsor's business plans and proposals, present and future products and policies, and methods of operation and prospects, are "Confidential Information" unless such information has already been fully disclosed in public literature, and the Company shall treat all such documents, material and information as confidential.

2.   Nondisclosure. (a) This Company agrees that it shall protect the Owner's/Sponsor's proprietary interest in the Confidential Information by, specifically but without limitation, limiting access to the Confidential Information to those employees, representatives and agents who (a) need to know the Confidential Information, (b) are informed by the Company of the confidential nature of the Confidential Information, and (c) agree to comply with the obligations and restrictions of this Agreement. The Company will not at any time, whether during or after its association with or engagement by the Owner/Sponsor, disclose to any person or entity or use any Confidential Information in any manner whatsoever or reproduce any Confidential Information in any form, in whole or in part, without the prior written consent of this Owner/Sponsor.

(b)   Notwithstanding the foregoing, if the Company or any of its employees, representatives or agents are required pursuant to applicable law, regulation or legal process to disclose any of the Confidential Information, it will notify the Owner/Sponsor promptly, prior to any such disclosure, so that a protective order or other appropriate remedy may be sought. If no such protective order is obtained, the Company will furnish only that portion of the Confidential Information, which is legally required, and will use all reasonable efforts to obtain reliable assurance that confidential treatment will be accorded the Confidential Information.

3.   Possession. This Company agrees that upon request by the Owner/Sponsor, and in any event, upon termination of its association with, or engagement by the Owner/Sponsor, the Company shall turn over to the Owner/Sponsor all documents, papers or other materials, whether recorded in printed form, electronically or otherwise, in its possession or under its control, which may contain or be derived from Confidential Information. The Company agrees that it shall have no proprietary interest in any work product developed or used by the Company and arising out of its engagement by this Owner/Sponsor.

4.   Remedy. The Company acknowledges that disclosure of any Confidential Information will give rise to irreparable injury to the Owner/Sponsor and/or customers of the Owner/Sponsor, and that money damages would be an inadequate remedy for any such breach. Accordingly, the Owner/Sponsor may seek and obtain injunctive relief against the breach or threatened breach or threatened breach of the foregoing, in addition to any other legal remedies, which may be available. The Company acknowledges and agrees that the restrictions contained herein are necessary for the protection of the Owner's/Sponsor's legitimate business interests and are reasonable in scope and content.

5. **Limited Enforcement**. The parties have carefully considered the restrictions herein and stipulated that they are fair and reasonable in light of all the facts and circumstances of the relationship between the parties; however, in the event a court should decline to enforce any restriction contained herein, that restriction shall be deemed to be modified to the maximum extent, which the court shall find enforceable. If any provision of this Agreement is found to be invalid by any court or tribunal, the invalidity of such provision shall not affect the validity of the remaining provisions hereof.

6. **Miscellaneous**. The provisions of this Agreement relating to confidentiality shall survive the termination of negotiations or engagement. All references herein to "the Owner/Sponsor" and the "Company" shall be deemed to include, without limitation, all subsidiaries, affiliates, shareholders, directors, officers, employees, agents, representatives, consultants and independent contractors. This Agreement shall be governed and construed according to the laws of the State of New York, without giving effect to the conflict of laws and principles thereof.

IN WITNESS WHEREOF, the Owner/Sponsor and the Company have duly executed this Agreement as of the later of the dates set forth below.

COMPANY

By: _____

Title: _____

OWNER/SPONSOR

By: _____

Title: _____

Note: Before using this template, all parties are urged to consult their attorney. This author makes no claim to its legality.

(Company Letterhead)

WAIVER OF LIEN TEMPLATE

Date _____

City_____ State _____

TO WHOM IT MAY CONCERN:

Whereas a contract has been made between the undersigned and *(corporate entity)*. To furnish labor _____

FOR A TOTAL AMOUNT OF _____ for the premises known as *(project location)*.

Now, therefore, the undersigned has been paid _____ on this contract to date. Upon final payment of _____, the undersigned does hereby waive and release any and all lien, and claim or right to lien on said above described premises under the laws relating to liens, on account of labor or materials, or both, furnished, or to be furnished, by the undersigned to aforesaid purchaser for said premises.

Company _____

Name/Title _____

BID FORM TEMPLATES

From: Name of Bidder _____

Address _____

_____

Telephone/Fax _____

To: Company Name
Address
Telephone Number:
Fax Number:

Attention _____

Project Name _____

A2.01 Base Bid: The undersigned having inspected the construction site and familiarized himself with all conditions likely to be encountered affecting the cost and schedule of work, having examined all of the contract documents, and having participated in an orientation pre-bid work session, hereby proposes to furnish all labor, materials, tools, equipment and services required to perform all of the work in strict accordance with the contract documents as prepared by:

_____

(Record Architect)

For the Base Bid Sum of:

$ _____

(Use figures)

_____

(Use words)

and if this proposal is accepted, will execute a form contract to this effect.

A2.02 BREAKDOWN OF BASE BID SUM:

A. The following breakdown of the base bid sum is submitted for use by the owner in evaluating this proposal and is not intended as a basis for establishing prices for additions to, or deletions from, the contract sum. The amounts shown include all labor, material, tools, equipment and services required for furnishing and installing each of the stated items of work, and include all other expenses, overhead and profit, properly distributed.

CSI CODE	DESCRIPTION	OTHER REF.	AMOUNT
	NEGOTIATED FEES		
	Contractor's Fee (including home office overhead)		
	General Conditions		
	Insurance		
	Bond Premium(s)		
	DIVISION 2 - SITEWORK		
02010	Subsurface Investigation		
02050	Demolition		
02100	Site Preparation		
02140	Dewatering		
02150	Shoring and Underpinning		
02160	Excavation Support Systems		
02170	Cofferdams		
02200	Earthwork		
02300	Tunneling		
02350	Piles and Caissons		
02500	Paving and Surfacing		
02600	Utility Piping Materials		
02660	Water Distribution		
02680	Fuel and Steam Distribution		
02700	Sewerage and Drainage		
02760	Restoration of Underground Pipe		
02780	Power and Communications		
02900	Landscaping and Signage		
	DIVISION 3 - CONCRETE		
03100	Concrete Formwork		
03200	Concrete Reinforcement		
03250	Concrete Accessories		
03300	Cast-In-Place Concrete		
03370	Concrete Curing		
03400	Pre-Cast Concrete		
03500	Cementitious Decks and Toppings		

CSI CODE	DESCRIPTION	OTHER REF.	AMOUNT
DIVISION 3 - CONCRETE			
03600	Grout		
03700	Concrete Restoration and Cleaning		
03800	Ready-Mix Concrete		
DIVISION 4 - MASONRY			
04100	Mortar and Masonry Grout		
04150	Masonry Accessories		
04200	Unit Masonry		
04400	Stone		
04500	Masonry Restoration and Cleaning		
04550	Refractories		
04600	Corrosion Resistant Masonry		
04700	Simulated Masonry		
DIVISION 5 - METALS			
05010	Metal Materials		
05030	Metal Coatings		
05050	Metal Fastening		
05100	Structural Metal Framing		
05200	Metal Joists		
05300	Metal Decking		
05400	Cold-Formed Metal Framing		
05500	Metal Fabrications		
05580	Sheet Metal Fabrications		
05700	Ornamental Metal		
05800	Expansion Control		
DIVISION 6 – WOOD AND PLASTICS			
06050	Fasteners and Adhesives		
06100	Rough Carpentry		
06130	Heavy Timber Construction		
06150	Wood and Metal Systems		
06170	Prefabricated Structural Wood		
06200	Finish Carpentry		
06300	Wood Treatment		
06400	Architectural Woodwork		
06500	Structural Plastics		
06600	Plastic Fabrications		
06650	Solid Polymer Fabrications		
DIVISION 7 – THERMAL AND MOISTURE PROTECTION			
07100	Waterproofing		
07150	Damp Proofing		
07190	Vapor Retardants		
07195	Air Barriers		

CSI CODE	DESCRIPTION	OTHER REF.	AMOUNT
DIVISION 7 – THERMAL AND MOISTURE PROTECTION			
07200	Interior Insulation		
07240	Exterior Insulation and Finish Systems		
07250	Fireproofing		
07270	Fire Stopping		
07300	Shingles and Roofing		
07400	Manufactured Roofing and Siding		
07480	Exterior Wall Assemblies		
07500	Membrane Roofing		
07570	Traffic Coatings		
07600	Flashing and Sheet Metal		
07700	Roof Specialties and Accessories		
07800	Skylights		
07900	Joint Sealers		
DIVISION 8 – DOORS AND WINDOWS			
08100	Metal Doors and Frames		
08200	Wood and Plastic Doors		
08250	Door Opening Assemblies		
08300	Special Doors		
08400	Entrances and Storefronts		
08500	Metal Windows		
08600	Wood and Plastic Windows		
08650	Special Windows		
08700	Hardware		
08800	Glazing		
08810	Glass		
08840	Plastic Glazing		
08900	Glazed Curtain Walls		
09100	Metal Support Systems		
09200	Lath and Plaster		
09250	Gypsum Board		
09300	Tile		
DIVISION 9 – FINISHES			
09400	Terrazzo		
09450	Stone Facing		
09500	Acoustical Treatment		
09540	Special Wall Surfaces		
09545	Special Ceiling Surfaces		
09550	Wood Flooring		
09600	Stone Flooring		
09630	Unit Masonry Flooring		
09650	Resilient Flooring		
09680	Carpet		
09700	Special Flooring		

CSI CODE	DESCRIPTION	OTHER REF.	AMOUNT
DIVISION 9 – FINISHES			
09780	Floor Treatment		
09800	Special Coatings		
09900	Painting		
09950	Wall Coverings		
DIVISION 10 - SPECIALTIES			
10100	Visual Display Boards		
10150	Compartments and Cubicles		
10200	Louvers and Vents		
10240	Grilles and Screens		
10260	Wall and Corner Guards		
10270	Access Flooring		
10290	Pest Control		
10300	Fireplaces and Stoves		
10340	Manufactured Exterior Specialties		
10350	Flagpoles		
10400	Identifying Devices		
10500	Lockers		
10520	Fire Protection Specialties		
10600	Partitions		
10650	Operable Partitions		
10670	Storage Shelving		
DIVISION 10 - SPECIALITIES			
10700	Exterior Protection Devices		
10750	Telephone Specialties		
10800	Toilet and Bath Accessories		
10900	Wardrobe and Closet Specialties		
DIVISION 11 - EQUIPMENT			
11010	Maintenance Equipment		
11020	Security and Vault Equipment		
11060	Theater and Stage Equipment		
11120	Vending Equipment		
11130	Audio-Visual Equipment		
11160	Loading Dock Equipment		
11170	Solid Waste Handling Equipment		
11200	Water Supply and Treatment Equipment		
11280	Hydraulic Gates and Valves		
11400	Food Service Equipment		
11460	Unit Kitchens		
11470	Darkroom Equipment		

CSI CODE	DESCRIPTION	OTHER REF.	AMOUNT
DIVISION 11 - EQUIPMENT			
11480	Athletic, Recreational and Therapeutic Equipment		
11500	Industrial and Process Equipment		
11680	Office Equipment		

CSI CODE	DESCRIPTION	OTHER REF.	AMOUNT
DIVISION 12 - FURNISHINGS			
12050	Fabrics		
12100	Artwork (Install Only)		
12300	Manufactured Casework (Install Only)		
12500	Window Treatment		
12600	Furniture and Accessories (Install Only)		
12670	Rugs and Mats		
12700	Multiple Seating		
12800	Interior Plants and Planters (Install Only)		
DIVISION 13 – SPECIAL CONSTRUCTION			
13400	Industrial and Process Control Systems		
DIVISION 14 – CONVEYING SYSTEMS			
14100	Dumbwaiters		
DIVISION 14 – CONVEYING SYSTEMS			
14200	Elevators		
14300	Escalators and Moving Walks		
14400	Lifts		
14500	Material Handling Systems		
14600	Hoists and Cranes		
14800	Scaffolding (See GC Item)		
DIVISION 15 – MECHANICAL SYSTEMS			
15050	Basic Mechanical Materials and Methods		
15250	Mechanical Insulation		
15300	Fire Protection		
15400	Plumbing		
15500	Heating, Ventilating and Air Conditioning		
15550	Heat Generation		
15650	Refrigeration		
15750	Heat Transfer		
15850	Air Handling		
15880	Air Distribution		
15950	Controls		
15990	Testing, Adjusting and Balancing		

CSI CODE	DESCRIPTION	OTHER REF.	AMOUNT
DIVISION 16 – ELECTRICAL			
16050	Basic Mechanical Materials and Methods		
16200	Power Generation Built-Up Systems		
16400	Service and Distribution		
16500	Lighting (Including Fixtures and Installation)		
16600	Special Systems		
16700	Communications		
16850	Electric Resistance Heating		
16900	Controls		
16960	Testing		
	TOTAL AMOUNT (must equal Base Bid Sum)		

A2.04 UNIT PRICE SCHEDULE

A.   All relevant unit costs shall be filled out in full detail.

SECTION 1 - ARCHITECTURAL	
Stone Flooring	/SF
Carpet	/SF
Resilient Flooring	/SF
Hollow Metal Door	/Unit
Wood Door + Frame	/Unit
Steel Partitions	/Unit
Hardware	/Unit
Type:	
Drywall/Plaster	/SF
Painting	/SF
Acoustic Ceiling/Lay-In Tile	/SF
Millwork	/Unit
Type:	

SECTION 2 – MECHANICAL (INSTALLED)	
Sheet Metal Ductwork	/LBS
Duct Insulation	/SF
Internal Lining	/SF
Diffusers	/Each
Grilles	/Each
Braised Condenser Water Copper Pipe	/LF
Soldered Condensate Drain Piping	/LF
Piping Insulation	/LF
SECTION 3 – SPRINKLER (INSTALLED)	
Relocate existing sprinkler head and piping	/Each
Install new sprinkler head and piping	/Each
SECTION 4 – ELECTRICAL (INSTALLED)	
Light Fixture	/Each
Type:	
SECTION 4 – ELECTRICAL (INSTALLED)	
Emergency Light	/Each
Exit Light	/Each
Lutron Lighting Switch	/
Time Clock	/
Subcontractor Markup	/
SECTION 5 – POWER/COMMUNICATION (INSTALLED)	
Homerun (3#12+G+IG)	/LFT
Double Duplex Receptacles (1)Reg+(1)IG	Wall
Double Duplex Receptacles (1)Reg+(1)IG	Floor
Voice/Data Outlet	/Each
Furniture System	
Power Feed	Each
Duplex Receptacle	Each
Core Drill for Floor Cell Access	Each
Voice/Data Cable	/LFT

A2.05 ALTERNATES

	ADD	DEDUCT
ALTERNATE 1		
ALTERNATE 2		
ALTERNATE 3		
ALTERNATE 4		

A2.06   CHANGES IN THE WORK

A.   The undersigned proposes that for changes in the work, subject to the applicable provisions of Article 7 of the General Conditions, the overhead and profit percentages to be added to the itemized estimates of cost will be the following:

	%	
CHANGE ORDER MARKUP/ CREDITS		

It is understood that changes in the work include all costs for supervision, engineering, office expense and any other general expense associated with and/ or resulting from changes in the work.

A2.07 ADDENDA: The undersigned acknowledges receipt of Addenda, as listed below, and represents that any additions or modifications to, or deletions from, the work called for in these Addenda, are included in the Base Bid Sum, and unit prices, if affected thereby.

ADDENDUM NO.   DATED

————————————   ————————
————————————   ————————
————————————   ————————
————————————   ————————
————————————   ————————
————————————   ————————
————————————   ————————

If no Addenda have been received, write in "NONE."

A2.08 TAXES

A.   It is agreed that all federal, state and local taxes, including, but not limited to, sales, use and excise tax that may be imposed on materials or services provided under this proposal are included in the Base Bid.

A2.09 SCHEDULE

A.   The undersigned understands that early completion is important to the owner and that the schedule will be considered in evaluation of proposals. Therefore, time being of the essence the undersigned proposes to perform the work in accordance with the following schedule of calendar days (all calendar days are measured from the date of award of the contract).

Calendar Days:

• Start of work

————————————————————————————————

- Substantially complete all work

- Final completion of all work

B.   The undersigned has enclosed herewith a construction schedule as required by instructions to bidders, which substantiates that the work can be accomplished within the scheduled dates quoted above.

A2.10 ASSIGNMENT: The undersigned represents that no assignment, sublease or transfer of all, or any part of, his interest in the proposal has been made, or will be made, without the written consent of the owner.

A2.11 BIDDING DOCUMENTS: The undersigned acknowledges receipt of the following bidding documents and affirms that all costs associated therewith are included in the applicable base bid sum and unit prices, including information transmitted at an initial orientation session.

AIA CONTRACT DOCUMENT CHECKLIST

Document Title	Document Number
• Bid Form	
• Building Rules and Regulations for Tenant Alterations	
• Standard Form of Agreement Between Owner and Contractor (Construction Contract)	A101
• General Conditions of the Contract for Construction	A201
• List of Subcontractors	G805
• Application and Certificate for Payment	G702
• Continuation Sheet	G703
• Shop Drawing and Sample Record	G712
• Certificate of Substantial Completion	G704
• Contractor's Affidavit of Release of Liens	G706A

A2.12 PROPOSED SUBCONTRACTORS AND MAJOR SUPPLIERS: The undersigned agrees, if awarded this contract by the owner, to promptly submit in writing a list of all prospective subcontractors and major material suppliers.

A2.13 DECLARATION: The undersigned declares, by executing the proposal, that:

- This proposal shall remain valid for acceptance by owner for a period not less than thirty (30) days from the bid due date.

- All requirements concerning licensing and all other local, state and national laws have been, or will be, complied with and that no legal requirements will be violated in the execution of the work.

- No person or persons or company other than the firm listed below or otherwise indicated hereinafter shall have any interest whatsoever in the proposal or the contract that may be entered into as a result thereof. This proposal is submitted in good faith, without collusion or fraud.

- The person or persons signing this proposal is/are fully authorized to sign on behalf of the conditions and provisions thereof.

SUBMITTED:

_____

(Name of Firm)

_____

(Address)

_____

_____

(Date)

OPERATING AS (Complete, and strike out words that do no apply)
A CORPORATION UNDER THE LAWS OF THE STATE OF:

_____

A PARTNERSHIP (GIVE FULL NAMES OF ALL PARTNERS):

_____

Enclosed: One (1) Construction Schedule

RECENTLY PUBLISHED

(OR TO BE PUBLISHED) WRITINGS,

USABLE AS

AN INFORMATION TOOL

OR A TRAINING GUIDE

CONSTRUCTION COST REDUCTION: WHAT'S IN A NAME?

By Adek Apfelbaum

Published as a part of *Cost Engineering,* Vol. 38/No. 4, copyright © April 1996.

The United States Army Corps of Engineers does everything "Texas style": big and bold. Thus, when on several major projects in the Middle East, the Corps renamed their estimating departments "EACC sections (engineering analysis and cost control)," and thereafter renamed all estimating departments "cost engineering branches." They may have stumbled upon something bigger than a bold name. Construction has always been plagued by a lack of control and overspending. Change orders (ECPs in the U. S. Army Corps of Engineers' lingo) have traditionally upset schedules and budgets usually in one direction—up! In turn, this created a need for value engineering (VE) to bring initial costs down or more in line with the functional requirements of a given project. Unfortunately, cost reduction fell on the shoulders of classical value engineering, a system that was devised for manufacturing and practiced mostly by industrial engineers as certified value specialists (CVSs). Their approach is to look at construction the way one looks at manufacturing: building under controlled conditions in a controlled environment. Although value engineering has been practiced in the construction industry for over thirty years, little progress has been made to eliminate super-high construction costs. Even less progress has been made to eliminate cost overruns. Under the current CVS-supervised cost reduction programs (still used by some public agencies), it is totally acceptable for the CVS to be (reduced to) a collector of suggestions from those who tend to create the problems in the first place, rather than being treated as expert enough to offer alternate solutions. This writer believes it is this topsy-turvy approach that renders value engineering impotent in controlling construction costs; only very obvious over-design can be eliminated under this format.

Construction costs are a part of market costs, estimating judgments and field conditions. Only someone who has been involved in all of the areas can see the not-so-obvious advantages of changing details or procedures to create immediate savings and savings during the life of the project (life-cycle costs). Savings cannot be visualized nor suggested by the narrow view of design/construction practitioners such as architectural dreamers or mechanical designers. Savings must be visualized by someone who sees the whole cost picture—someone who can predict the interrelated costs, including costs for indirect items such as time loss, site accessibility, constructability and other general condition items. A senior estimator has an instinctive feel for costs. With some training and standardization, he can use this acquired know-how to produce cost reductions based on past experience and predetermined principles. Further, since change orders also require an estimator's evaluations, he is in the perfect position to follow through, establish a budget, hold the team to this goal and control and follow the increases (those necessary and those snuck in for various reasons).

Therefore, this is a plea for giving estimators more control over their work by advancing senior estimators to handle cost reductions and cost control. Such estimators should be renamed cost reduction engineering analysts or CREAS. Practicing value engineers or certified value specialists are needed and should continue to practice as they wish, but mostly in manufacturing processes. For our industry, CREAs would be better equipped and more appropriately named. Cost reduction goes hand in hand with cost control, which goes hand in hand with estimating, which should be the cost reduction, engineering analysts' domain.

CONSTRUCTION MANAGERS AND ESTIMATORS: A VERY TENUOUS RELATIONSHIP

By Adek Apfelbaum

Published as a part of *Cost Engineering,* Vol. 38/No. 4, copyright © April 1996.

Estimators/cost engineers are generally viewed as an appendage to any organization. They are stuck in a corner and expected to crunch numbers, interpret cost and mainly keep the firm or its project managers from under-pricing initial costs/bids and changes in scope. The role of an estimating department is clear in a GC (general contractor) or subcontractor environment. However, not so clear is how estimators function in a CM-GMP (construction manager-guaranteed maximum price) environment. During the initial stages, when the final design is still unresolved, when the GMP has not yet been fixed, estimators function basically the same under a GC or CM format. Both conditions call for a conceptual budget, a wish list/program analyses. As the design approaches completion, CM and GC needs grow further apart. While a GC can hide behind a not-so-defined scope (and take exceptions and exclusions), a CM must guarantee the maximum cost before the final design has been created and before the final scope has been fixed. Under these conditions, wish list analysis or conceptual estimating (macro-estimating) is the preferred tool. The same approach under a general contractor format may be viewed as guessing ("guesstimating") or other forms of inaccurate evaluations. Estimates, as used by construction managers, are still greatly misunderstood, especially when change orders are involved. A general contractor's or subcontractor's estimate must cover as many possibilities as can be determined, including as many unknowns that can be predicted and as much of what was left out from the original evaluation of the subsystem being estimated. On the other hand, a CM must look out for the owner's interest, offering a specific contractor only what the subsystem is worth. Any side issues, such as the same contractor having been unduly disadvantaged before, have no validity in any given evaluation of extras/credits. A CM's estimator should know this and, therefore, must strive to produce intrinsic macro-estimates as a basis for comparison between what (how much) the contractor wants and what he deserves. A CM should not expect his estimator to be more than a value barometer. Further, a CM's project manager should understand that a contractor's evaluations may be tainted and are not necessarily (and automatically) more correct than those that were generated in-house. Also, in-house estimates need not be lower than the contractor's to be a good tool for negotiating or upgrading contractual scope and price. In his assigned capacity, a project manager uses an estimator's "macro" evaluation as a negotiating wedge and an x-ray view of what the contractor is trying to achieve. For the reasons cited, the CM's estimator should not be required to adjust his estimate to accommodate the contractor's. In short, a CM's estimator needs to be a visionary, a conceptual macro-assessor and one who sees the whole picture, but is not burdened with insignificant facts or costs. A CM must be aware of these subtle realities. A CM should also be aware that his needs for estimating, more often than not, differ from those of a GC. By necessity, a GC's estimate must be a tool to create a safe bottom line; a CM's estimate is an assessment of the intrinsic worth to be used as a tool for negotiating with contractors, keeping them in line and "honest."

ARCHITECTURE
AND COST AWARENESS

By Adek Apfelbaum

Submitted for publication in *Architecture*

Cost cutting has always been considered crude and working against esthetics. Architects very often want no part of it. During the various building-boom periods, any design, no matter how uneconomical, was usually funded and built. Now, when funds are scarce, availability of architectural commissions seem to depend greatly on designs being economical or fundable. In short, commissions are mostly driven by economics and functionality.

Good architecture has always been costly beyond its intrinsic value. As a result, institutional and private owners/sponsors turned to VE (value engineering), a system devised for industrial cost efficiency, to halt the accelerated spiraling of construction costs. Governmental agencies even offered contractors a portion of the savings resulting from value-engineered solutions the contractor may suggest. Thus, value engineering became the buzzword during the past twenty years. Despite all the publicity it received, value engineering produced no slowdown of the constant upward movement of cost. Value engineering (through their practitioners, CVSs, certified value specialists) is driven by the formula $V + F/C$: increase the function (F) or decrease the cost (C) and value (V) is increased. Unfortunately, CVSs have been more successful in increasing "F" than decreasing "C." This drove some estimators and construction-orientated managers to the cost engineering aspect in order to find answers from the design/construction approach. Thus, "applied VE" and "intuitive VE" were born. Many construction managers and construction service providers still felt obligated to "dismantle" design enhancements for the sake of economics. Architects became wary of the term "cost reduction." The term became synonymous with design meddling (i.e., cost cutting equals design cutting).

From this chaos order seems to be emerging. Some senior estimators, who have respect for the esthetics of design, recently seized this opportunity for the development of a cost reduction procedure, which is under the architects control and which is geared totally to the design/construction process. The formula $V = F/C$ is still used but with a variation: $V = F + E/C$ (where the "E" stands for esthetics). The variable is "C" and "F + E" is left totally to the architect. Further, the services are offered by the designer, who acquires the part-time use of a CREA (cost reduction engineering analyst), a generalist who works at the designer's discretion under a system called ID/C.

There seems to be an undeclared battle by the newly created CREAS, CCCRs or the more advance CCCEs (certified construction cost engineers) for the takeover of construction cost reduction. It is likely that CREAS/CCCEs, who may soon be certified by the A.A.C.E. (Association for the Advancement of Construction Cost Engineering), will win this conflict. Their approach is more aligned with the design team, and working under its direction (and from its office) can do little design damage. CREAs can make the design team more responsive to a client's wish list and the size of his money belt. War or peace, ID/C is here to stay. It is a logical outcome of a progression of events. The power to "cut and how to cut" has finally been returned to the creative force, the design team.

ABOUT THE AUTHOR

The author has "formulated" cost engineering as an advanced form of construction estimating. He is a graduate civil engineer with forty years of construction experience. He has taught, written about and practiced the art of cost engineering for the past thirty years. He has published numerous articles on the topic and has provided industry application formats for the American Army Corp of Engineers and more recently, Pradausa on its construction program in the U.S.A. He has formed The Society for the Advancement of Construction Cost Engineering (SACCE) with a basic nucleus of twenty qualified members who will act as teachers, practitioners and preachers.

Printed in the United States
42002LVS00002B/38